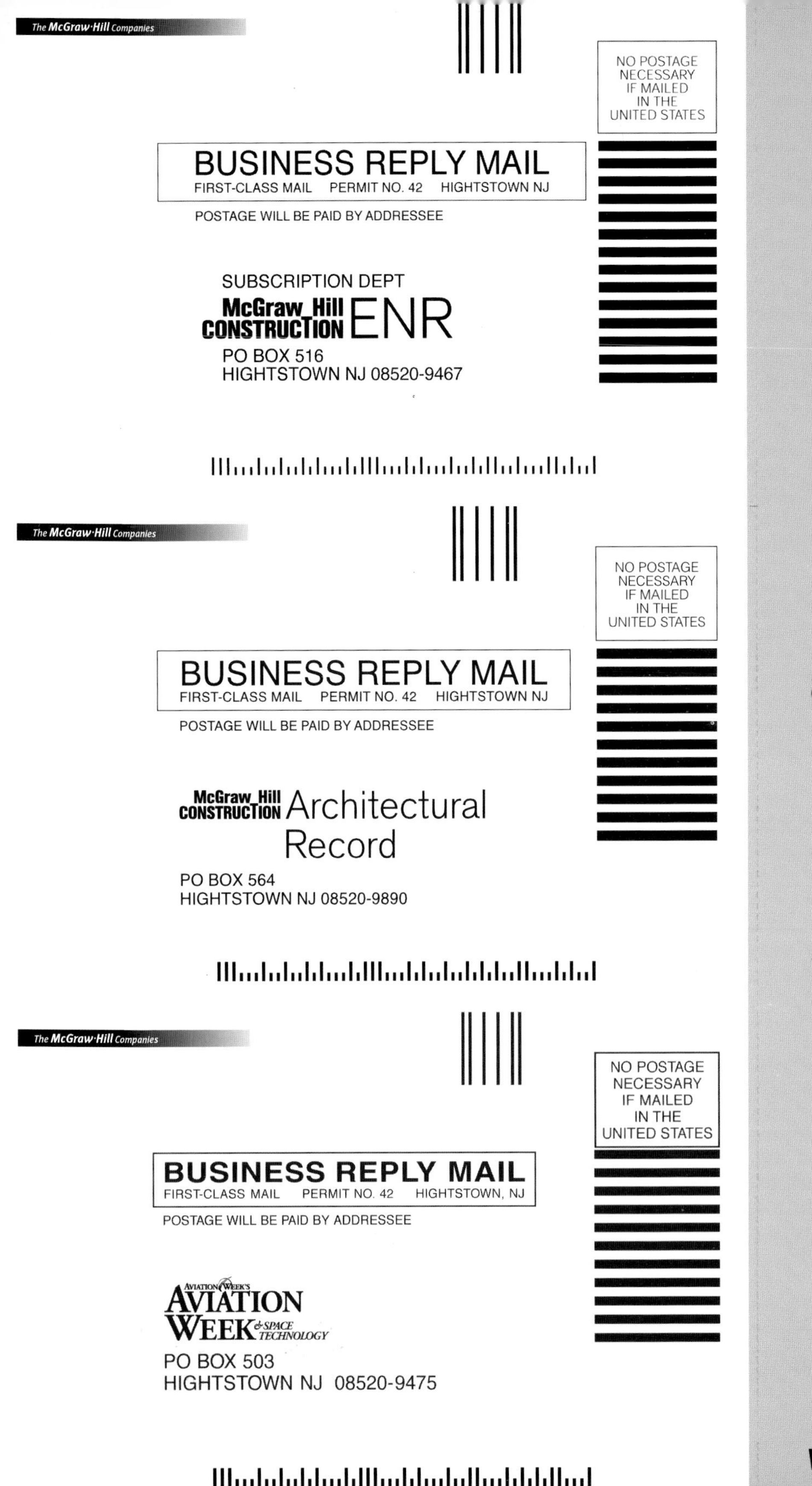

Stay on top of industry developments with the leading publications in your engineering field.

Today's economy requires a higher level of knowledge and adaptabilty—increase yours with **McGraw-Hill Construction and Aviation Week & Space Technology:** *Engineering News-Record, Architectural Record, and Aviation Week.*

www.construction.com

www.AviationNow.com/awst

Teamwork and Project Management

Karl A. Smith
University of Minnesota

Second Edition

Boston Burr Ridge, IL Dubuque, IA Madison, WI New York San Francisco St. Louis
Bangkok Bogotá Caracas Kuala Lumpur Lisbon London Madrid Mexico City
Milan Montreal New Delhi Santiago Seoul Singapore Sydney Taipei Toronto

TEAMWORK AND PROJECT MANAGEMENT, SECOND EDITION

Published by McGraw-Hill, a business unit of The McGraw-Hill Companies, Inc., 1221 Avenue of the Americas, New York, NY, 10020.
Previously published under the title *Project Management and Teamwork* by The McGraw-Hill Companies, Inc.

This book is printed on acid-free paper.

1 2 3 4 5 6 7 8 9 0 QPF/QPF 0 9 8 7 6 5 4 3

ISBN 0-07-248312-1

Publisher: *Elizabeth A. Jones*
Senior sponsoring editor: *Carlise Paulson*
Marketing manager: *Sarah Martin*
Project manager: *Sheila M. Frank*
Senior production supervisor: *Laura Fuller*
Senior coordinator of freelance design: *Michelle D. Whitaker*
Cover images: © *Photodisc, Inc.*
Compositor: *ElectraGraphics, Inc.*
Typeface: *10/12 Palatino*
Printer: *Quebecor World Fairfield, PA*

Library of Congress Cataloging-in-Publication Data

Smith, Karl A.
Teamwork and project management / Karl A. Smith.—2nd ed.
p. cm.
Rev. ed. of: Project management and teamwork. © 2000.
Includes bibliographical references and index.
ISBN 0-07-248312-1
1. Engineering management. 2. Teams in the workplace. 3. Project management.
I. Smith, Karl A. Project management and teamwork. II. Title.

TA190.S63 2004
658.4'04—dc21

2003048799
CIP

www.mhhe.com

Contents

Preface

When McGraw-Hill invited me to write a module on project management and teamwork for their BEST series, I thought, What a terrific idea! I had been teaching project management and teamwork courses for seniors in engineering; graduate students in professional master's programs, especially at the University of Minnesota's Center for the Development of Technological Leadership; and participants in short courses in the University of Minnesota Executive Development Program, government agencies, and private companies. It would not have occurred to me to write a book for first-year students. I immediately embraced the idea and started work.

I've been teaching a course for first-year students at the University of Minnesota for more than 20 years. It has evolved into a course titled How to Model It: Building Models to Solve Engineering Problems, which I have been teaching with colleagues and undergraduate student teaching assistants for the past 10 years. We also wrote a book to accompany the course—*How to Model it: Problem Solving for the Computer Age* (Starfield, Smith, and Bleloch, 1994). Since this course makes extensive use of project teams, I know that a book on project management and teamwork is needed.

Teamwork and projects are at the heart of the approach I use in teaching students at all levels, including participants in faculty development workshops. I've learned that it isn't easy for students to work effectively in project teams or for faculty to organize and manage them, but the potential for extraordinary work from teams makes it worth the effort. Also, projects and teamwork are a central part of engineering work in the world outside the classroom.

The first part of this book summarizes the context of engineering and stresses the importance of teamwork. The middle part focuses on the nature of projects and the project managers' role. The last part emphasizes the particulars on scheduling, monitoring, and documentation. Overall, my goals for readers of *Project Management and Teamwork* are the following:

- To understand the dynamics of team development and interpersonal problem solving.
- To identify strategies for accelerating the development of true team effectiveness.
- To understand the critical dimensions of project scope, time, and cost management.
- To understand critical technical competencies in project management.
- To explore a variety of "best practices" including anticipating, preventing, and overcoming barriers to project success.

As you engage with this book, be sure to continually reflect on what you're learning and how you can apply it to the projects and teams you work on each day, in classes, on the job, and in social, professional, and community organizations. An important key to success in projects and teams is to routinely work at a "meta level." That means you are simultaneously thinking about the task and how well the team is working. Talk with others about how the projects and teams you're involved with are going, share successes and insights, and work together to identify and solve team problems. The personal story in the accompanying box describes some of the questions I've grappled with and how I got interested in this project. I encourage you to develop your own stories as you work your way through this book.

One of the messages of the story in the box is the importance of checking a variety of resources to help formulate and solve the problems you encounter. Another message is that, although engineers spend some of their time working alone, engineering is not individual, isolated work. Collaborative problem solving and teamwork are central to engineering. Engineers must learn to solve problems by themselves, of course, but they must also learn to work collaboratively to effectively solve the other 95 percent of the problems they will face as professional engineers. There may be a tendency to think that this 95 percent—this asking questions and searching other sources for the solution—is either trivial or else unrelated to engineering. However, working with others to formulate and solve problems and accomplish joint tasks is critical to success in engineering.

Personal Story

I have been involved in engineering, as a student and as a professional, for over 30 years. Frequently I have grappled with the questions, What is the engineering method? Is it applied science? Is it design? As a professor I have struggled with the question, What should my students learn and how should they learn it? These concerns prompted me to address the question, What is the nature of engineering expertise and how can it be developed effectively?

A study conducted by one of my colleagues (Johnson, 1982) provides valuable insight into the activities of engineers. My colleague was hired to collect protocol from engineering experts while they solved difficult problems. Working with a team of professors, he developed a set of difficult and interesting problems, which he took to chief engineers in large companies. In case after case the following scenario was repeated.

The engineer would read the problem and say, "This is an interesting problem."

My colleague would ask, "How would you solve it?"

The engineer would say, "I'd check the engineers on the floor to see if any of them had solved it."

In response, my colleague would say, "Suppose that didn't work."

"I'd assign the problem to one of my engineers to check the literature to see if a solution was available in the literature."

"Suppose that didn't work," retorted my colleague.

"Well, then I'd call my friends in other companies to see if any of them had solved it."

Again my colleague would say, "Suppose that didn't work."

"Then I'd call some vendors to see if any of them had a solution."

My colleague, growing impatient at not hearing a problem solution, would say, "Suppose that didn't work."

At some stage in this interchange, the engineer would say, "Well, gee, I guess I'd have to solve it myself."

To which my colleague would reply, "What percentage of the problems you encounter fall into this category?"

Engineer after engineer replied, "About five percent"!

Acknowledgments

Many people deserve credit for guidance in this project. Michael B. Mahler, a graduate student in civil engineering at the University of Minnesota, with whom I've taught and worked on project management for many years, provided enormous insight into the process of what will work for students and was a source of constant support and encouragement. Robert C. Johns co-taught the project management course with me at Minnesota and provided lots of good ideas. Anthony M. Starfield, co-creator of the first-year course, How to Model It, and co-author of the book by the same title encouraged me to use the questioning format of the *How to Model It* book to engage the reader. The five manuscript reviewers provided terrific assistance. Holly Stark and Eric Munson of McGraw-Hill, and Byron Gottfried, Consulting Editor, initiated the idea and provided guidance throughout. A special note of thanks to my daughters, Riawa and Sharla Smith, who helped with the editing and graphics.

A special acknowledgment to Michigan State University, which provided me with a wonderful place to work on this project during my sabbatical leave. Another goes to David and Roger Johnson (whose cooperative learning model provides the theoretical basis for this book) for their great ideas, generosity, and steadfast support.

Most of all I thank the hundreds of students who learned from and with me in project management courses for their patience, perseverance, wonderful suggestions and ideas, and interest and enthusiasm in project management and teamwork.

Comments and Suggestions

Please send your comments and suggestions to me at ksmith@umn.edu.

Preface to the Second Edition

Welcome to the second edition, now retitled *Teamwork and Project Management.* Many things have changed since I wrote the first edition in 1998–1999. Teamwork has received increased emphasis from ABET and from employers, the world has grown smaller and our sense of interdependence has greatly increased, the importance of professional responsibility and ethics has magnified, projects are becoming much more common. Because teamwork and projects are prevalent in engineering in business, industry, and government, they are also becoming common in engineering classes. In addition to the importance of teamwork in the profession, teams are used in classes because students working in well-structured teams learn more, remember it longer, and develop superior problem-solving skills compared with students working individually. All these changes increase the importance of learning (and practicing) the concepts, principles, and heuristics in this book.

My civil engineering project management course is overflowing with students from across the Institute of Technology at the University of Minnesota. They apparently are voting with their feet, recognizing the importance of teamwork and project management skills. I really appreciate their enthusiasm. The teaching team has grown considerably and now includes several graduate students. The current teaching team includes Brandon Pierce, Connie Kampf, and Lori Engstrom. They have been wonderful in helping revise the course, and therefore have had lots of influence on this book. Two adjunct faculty, Tim Eiler and Randy Carlson, will start teaching the course this year, and I suspect the next iteration of this book will include many of their ideas.

The reviewers and editors made many wonderful suggestions for improving the book, many of which I've incorporated in this edition. The most notable is probably the title change from *Project Management and Teamwork* to *Teamwork and Project Management,* which was suggested by Kelly Lowery and John Griffin.

Project Management and Teamwork was designed for first-year students, but it has been used by other students, especially those in senior-level capstone design courses. *Teamwork and Project Management* is still designed to be accessible to first-year students, but will be applicable for upper-division students who haven't had an opportunity to focus on teamwork and project management skills in earlier courses and programs.

Chapter 1, which is an introduction and overview, was extensively revised. Chapters 2 and 3, the teamwork chapters, were updated and expanded. Chapters 4 and 5, on project management basics, were rearranged and new material on scoping projects was added, based on new developments and the importance of planning. Errors were corrected in Chapter 6, but that chapter wasn't changed much otherwise. The remainder of the book was updated.

Tom Peters wrote in his book *The Project 50,* "In the new economy, all work is project work." My intention is that this book will help prepare you to work in the new economy. Good teamwork and project work to you!

References

Johnson, P. E. 1982. Personal communication.

Peters, Tom. 1999. *The project 50: Fifty ways to transform every "task" into a project that matters.* New York: Knopf.

Starfield, Anthony M., Karl A. Smith, and Andrew L. Bleloch. 1994. *How to model it: Problem solving for the computer age.* Edina, MN: Interaction Book.

CHAPTER 1

Teamwork and Project Management in Engineering

Teamwork and Project Management is designed to help you prepare for professional practice in the new economy. Teamwork is receiving increased emphasis from employers. The world has gotten smaller and our sense of interdependence has greatly increased, the importance of professional responsibility and ethics has magnified, and projects (and project-type organizations) are becoming much more common. All these changes highlight the importance of learning (and practicing) the concepts, principles, and heuristics in this book.

According to Thomas Friedman (2000), "the world is ten years old." Friedman's central notion is *globalization:* "the inexorable integration of markets, nation-states, and technologies to a degree never witnessed before—in a way that is enabling individuals, corporations and nation-states to reach around the world farther, faster, deeper and cheaper than ever before, and in a way that is enabling the world to reach into individuals, corporations, and nation-states farther, faster, deeper, and cheaper than ever before" (p. 9).

This is the world in which you'll be working. It is very different from the world I started working in as an engineer in 1969, but it is the world I try to cope with every semester with undergraduate students in civil engineering and graduate students in three professional masters programs in which I teach—Management of Technology, Manufacturing Systems Engineering, and Infrastructure Systems Engineering. The engineering graduates in these one-day-per-week, two-year programs are working full-time and most of the participants work globally.

The essence of the globalization economy (according to Surowiecki, 1997) is this notion: "Innovation replaces tradition. The present—or perhaps the future—replaces the past." The importance of innovation and creativity in engineering is one of the new emphases in this edition.

As we start this journey together, I offer you some suggestions that I think will help you get the most from this book. The essence of the suggestions are *activity, reflection,* and *collaboration.* First, I encourage you to engage in the activities, especially the exercises in the book, as they will help connect you with the material and its real-world applications. Second, periodically throughout the book I'll ask you to stop and reflect. I encourage you to take advantage of the opportunity. My goal is to give you a chance to describe what you already know and to get you to think. Then when you read what I have to say about the topic, you'll have a basis for comparing and contrasting. Finally, I encourage you to collaborate with others. Working together is the norm in projects. Working together to learn the material in this book will make it easier, and very likely you'll remember it longer.

My goal for this chapter is to create a context for teamwork and project management in engineering. Let's start by exploring the nature of engineering. Before you read ahead for various answers to the question "What is engineering?" please complete the following reflection.

What Is Engineering?

> ☞ REFLECTION What is engineering? What does it mean to learn to engineer in school? What is your experience with engineering? Did you learn about engineering in high school? Do you have a brother or sister, mother or father, or other family relative or friend who is an engineer? Take a minute to reflect on where you learned about engineering and what your impressions of engineering are.
>
> What did you come up with?

Because there are few high school courses in engineering, most first-year students have not had much exposure to engineering. Yet we are surrounded by engineering accomplishments; they are so ubiquitous that we don't notice most of them. One of the foremost thinkers and writers on engineering, mechanical engineering professor Billy Koen, is noted for asking four probing questions of his audiences (Koen, 1984). The first is this:

1. Can you name one thing in the room in which you are sitting (excluding yourself, of course) that was not developed, produced, or delivered by an engineer?

Koen finds that the question is usually greeted with bewildered silence. I have posed Koen's questions to hundreds of first-year students, and they come up with some great suggestions: the air (but how does it get into the room?), dirt (trapped in people's shoes), electromagnetic radiation (but the lights generate much more than the background). Almost everything that we encounter was developed, produced, or delivered by engineers.

Here is Koen's second question:

2. Can you name a profession that is affecting your life more incisively than engineering?

Again, students name several professions but on reflection note that if it were not for engineering, politicians would have a difficult time spreading their ideas; doctors, without their tools, would be severely limited in what they could do; lawyers wouldn't have much to read; and so forth. Things such as telephones, computers, airplanes, and skyscrapers—which have enormous effects on our lives—are all products of engineering.

Koen's third question is this:

3. Since engineering is evidently very important, can you now define the engineering method for solving a problem?

Many students respond with a puzzled look, as if I am asking an unfair question. They note that they have a ready response to the question "What is the scientific method?" Students list things like "applied science," "problem solving," and "trial and error," but almost no one (over the 20 or so years that I've been asking this question) says "design."

If you were to ask practicing engineers what the engineering method is, they would likely respond "Engineering is design!" A group of national engineering leaders has said:

> Design in a major sense is the essence of engineering; it begins with the identification of a need and ends with a product or system in the hands of a user. It is primarily concerned with synthesis rather than the analysis which is central to engineering science. Design, above all else, distinguishes engineering from science. (Hancock, 1986)

We'll explore the concept of engineering design next—and save Koen's fourth and final question for the end of the chapter.

Engineering Design

If design is the essence of engineering, the next question is, What is design? What comes to mind when you consider the term *design?* Do you think of product design (such as automobiles), architectural design, set and costume design (as in theater), or interface design (as in computer)? Take a moment to collect your thoughts on design.

The Accreditation Board for Engineering and Technology (ABET), the group that accredits engineering programs, defined engineering design as "the process of devising a system, component or process to meet a desired need" (ABET, 2000).

Researchers who carefully observe the engineering design process are increasingly noting that it is quite different from the formal process typically described in textbooks. For example, Eugene Ferguson (1992, p. 32) writes:

> Those who observe the process of engineering design find that it is not a totally formal affair, and that drawings and specifications come into existence as a result of a social process. The various members of a design group can be expected to have divergent views of the most desirable ways to accomplish the design they are working on. As Louis Bucciarelli (1994), an engineering professor who has observed engineering designers at work, points out, informal negotiations, discussions, laughter, gossip, and banter among members of a design group often have a leavening effect on its outcome.

Recent work on engineering design indicates that design is a more social process than we once thought. Larry Leifer (1997) of the Stanford Center for Design Research claims that engineering design is "a social process that identifies a need, defines a problem, and specifies a plan that enables others to manufacture the solutions." Leifer's research shows that design is fundamentally a social activity. He describes practices such as "negotiating understanding," "conserving ambiguity," "tailoring engineering communications for recipients," and "manipulating mundane representations."

If design is the heart of engineering and design is a social process, then it follows that teamwork and project management are essential to engineering. Many problems with engineering result from poor team dynamics and inadequate project management.

Design team failure is usually due to failed team dynamics.
LARRY LEIFER
Director, Stanford Center for Design Research

A lot has been written about engineering and engineering design. Adams (1991), Hapgood (1992), and Ferguson (1992), for example, can give students considerable insight into engineering. One of the most interesting insights into engineering design was presented in the ABC News *Nightline* show documenting the design process at the product design firm IDEO ("The Deep Dive," July 13, 1999). David Kelly, head of IDEO, challenged the viewer: "look around—the only things not designed by humans are in nature." Five steps are key to IDEO's expertise in innovative design:

1. Understand the market/client/technology/constraints.
2. Observe real people in real situations.
3. Visualize new-to-the-world concepts and ultimate customers.
4. Evaluate and refine prototypes.
5. Implement new concepts for commercialization.

I hope you have an opportunity to view this show on video or DVD. Students I've shown it to exclaim, "I want to work at a place like that!"

Now that we've taken a glance at engineering and the role of design, let's turn to the role of teamwork and project management in engineering.

Employer's Checklist—Boeing Company

✔ A good grasp of these engineering fundamentals:
  Mathematics (including statistics)
  Physical and life sciences
  Information technology
✔ A good understanding of the design and manufacturing process (i.e., an understanding of engineering)
✔ A basic understanding of the context in which engineering is practiced, including:
  Economics and business practice
  History
  The environment
  Customer and societal needs
✔ A multidisciplinary systems perspective
✔ Good communication skills
  Written
  Verbal
  Graphic
  Listening
✔ High ethical standards
✔ An ability to think critically and creatively as well as independently and cooperatively
✔ Flexibility—an ability and the self-confidence to adapt to rapid/major change
✔ Curiosity and a lifelong desire to learn
✔ A profound understanding of the importance of teamwork

Source: Briefings ASEE Prism, December 1996, p. 11.

Teamwork and Engineering

How important is teamwork in the practice of engineering? Take a moment to reflect on your experiences and conversations on the importance and role of teamwork in engineering practice.

The quotes from Rockefeller and Welch (p. 25) stress the importance of teamwork from the perspective of a corporate chief executive officer (CEO), but what about its importance for engineering graduates?

Teamwork and project management are central to engineering. Learning how to organize and manage projects, and to participate effectively in project teams, not only will serve you well in engineering school, where there are lots of group projects, but also will be critical to your success as a professional engineer. The Boeing Company uses the checklist on page 4 of the characteristics they want in their employees when considering new applicants.

The Boeing Company checklist has been undergoing updates and refinements, and the following were added (or revised extensively) in a list titled "Desired Attributes of a Global Engineer":

- ✔ An awareness of the boundaries of one's knowledge, along with an appreciation for other areas of knowledge and their interrelatedness with one's own expertise
- ✔ An awareness and strong appreciation for other cultures and their diversity, their distinctiveness, and their inherent value
- ✔ A strong commitment to teamwork, including extensive experience with and understanding of team dynamics
- ✔ An ability to impart knowledge to others.

The emphasis on teamwork is not entirely new, as shown in the following 1988 list of skills employers want their employees to have.

What Employers Want

- Learning to learn
- Listening and oral communication
- Competence in reading, writing, and computation
- Adaptability: Creative thinking and problem solving
- Personal management: Self-esteem, goal setting/motivation, and personal/career development
- Group effectiveness: Interpersonal skills, negotiation, and teamwork
- Organizational effectiveness and leadership

Source: Workplace basics: The skills employers want. 1988. American Society for Training and Development and U.S. Department of Labor.

The importance of teamwork in business and industry is also embedded in the concepts of concurrent (or simultaneous) engineering and total quality management. The following quote elaborates on this point:

> In concurrent engineering (CE), the key ingredient is teamwork. People from many departments collaborate over the life of a product—from idea to obsolescence—to ensure that it reflects customers' needs and desires. . . . Since the very start of CE, product development must involve all parts of an organization,

> effective teamwork depends upon sharing ideas and goals beyond immediate assignments and departmental loyalties. Such behavior is not typically taught in the engineering schools of U.S. colleges and universities. For CE to succeed, teamwork and sharing must be valued just as highly as the traditional attributes of technical competence and creativity, and they must be rewarded by making them an integral part of the engineer's performance evaluation. (Shina, 1991, p. 23)

The increased emphasis on teamwork in engineering classes is partly due to the emphasis by employers, but it is also due to engineering education research on active and cooperative learning, and the emphasis of ABET. To maintain ABET accreditation, engineering departments must demonstrate that all of their graduates have the following eleven general skills and abilities (ABET, 2000):

1. An ability to apply knowledge of mathematics, science, and engineering
2. An ability to design and conduct experiments, as well as to analyze and interpret data
3. An ability to design a system, component, or process to meet desired needs
4. An ability to function on multidisciplinary teams
5. An ability to identify, formulate, and solve engineering problems
6. An understanding of professional and ethical responsibility
7. An ability to communicate effectively
8. The broad education necessary to understand the impact of engineering solutions in a global and societal context
9. A recognition of the need for, and an ability to engage in, lifelong learning
10. A knowledge of contemporary issues
11. An ability to use the techniques, skills, and modern engineering tools necessary for engineering practice

As you no doubt have recognized, a confluence of pressures emphasizes teamwork in engineering education and practice. We need to leave room for the "maverick," but most, if not all, engineering graduates need to develop skills for working cooperatively with others—as indicated by the lists of the top three engineering work activities.

Top Three Main Engineering Work Activities

Engineering Total	**Civil/Architectural**
• Design—36%	1. Management—45%
• Computer applications—31%	2. Design—39%
• Management—29%	3. Computer applications—20%

Source: Burton, L., Parker, L., & LeBold, W. 1998. U.S. engineering career trends. *ASEE Prism* 7(9), 18–21.

The full list of work activity reported by engineers is shown in the Table 1.1. Note that 66 percent mentioned design and 49 percent mentioned management.

Fundamental Tools for the Next Generation of Engineers and Project Managers

I've stressed the importance of teamwork for engineering education and practice, but teamwork isn't all that's needed. If engineers are going to become "the

Table 1.1 Rank Order of Work Activities, 1993

Activity	Percentage Mentioning
1. Design	66
2. Computer applications	58
3. Management	49
4. Development	47
5. Accounting, etc.	42
6. Applied research	39
7. Quality or productivity	33
8. Employee relations	23
9. Sales	20
10. Basic research	15
11. Production	14
12. Professional services	10
13. Other work activities	8
14. Teaching	8

Source: Burton, Parker, and LeBold, 1998, p. 19.

master integrators," as emphasized by Joe Bordona (1998), three additional tools are fundamental:

- Systems/systems thinking/systems engineering
- Models
- Quality (I defer this discussion to Chapter 7)

The Systems Approach

Employer checklists like Boeing's and the new ABET accreditation criteria emphasize systems and the systems approach.

A *system* is a whole that cannot be divided up into independent parts (Ackoff, 1994). Systems are made up of sets of components that work together for a

The Art and Practice of the Learning Organization

1. *Building shared vision.* The idea of *building* shared vision stresses that you never quite finish it—it's an ongoing process.
2. *Personal mastery.* Learning organizations must be fully committed to the development of each individual's personal mastery—each individual's capacity to create their life the way they truly want.
3. *Mental models.* Our vision of current reality has everything to do with the third discipline—*mental models*—because what we really have in our lives is constructions, internal pictures that we continually use to interpret and make sense out of the world.
4. *Team learning.* Individual learning, no matter how wonderful it is or how great it makes us feel, is fundamentally irrelevant to organizations, because virtually all important decisions occur in groups. The learning units of organizations are "teams," groups of people who need one another to act.
5. *Systems thinking.* The last discipline, the one that ties them all together, is *systems thinking.*

Source: Senge, 1993.

specified overall objective. The systems approach is simply a way of thinking about total systems and their components.

Five basic considerations must be kept in mind when thinking about the meaning of a system: (1) the total system's objectives and, more specifically, the performance measures of the whole system; (2) the system's environment: the fixed constraints; (3) the resources of the system; (4) the components of the system, their activities, goals, and measures of performance; and (5) the management of the system (Churchman, 1968).

Systems thinking is a discipline for seeing wholes—a framework for seeing interrelationships rather than things, for seeing patterns of change rather than static "snapshots." It is a set of principles and a set of specific tools and techniques (Senge, 1990). An implication of the systems approach is that it is important to get everybody involved to improve whole systems (Weisbord, 1987). The systems approach is commonly operationalized through learning organizations (see the box "The Art and Practice of the Learning Organization").

A systems theme is one of the integrating threads in this book. The concepts of systems and of the learning organization are important not only to your study of teamwork and project management but to many other things you will be studying in engineering. Here, for example, are eight principles for learning from Xerox (Jordon, 1996, p. 116):

1. Learning is fundamentally social.
2. Cracking the whip stifles learning.
3. Learning needs an environment that supports it.
4. Learning crosses hierarchical bounds.
5. Self-directed learning fuels the fire.
6. Learning by doing is more powerful than memorizing.
7. Failure to learn is often the fault of the system, not the people.
8. Sometimes the best learning is unlearning.

This list from Xerox indicates that the ideas in this book are important not only for your project work but also for your day-to-day work in engineering school.

Model and Modeling

Modeling in its broadest sense is the cost-effective use of something in place of something else for some cognitive purpose (Rothenberg, 1989). A model represents reality for the given purpose; the model is an abstraction of reality, however, in the sense that it cannot represent all aspects of reality. Models are characterized by three essential attributes:

1. *Reference:* A model is *of* something (its *referent*).
2. *Purpose:* A model has an intended cognitive *purpose* with respect to its referent.
3. *Cost-effectiveness:* A model is more *cost-effective* to use for this purpose than the referent itself would be.

I often give students this problem to help them learn about these attributes of modeling:

Determine the maximum number of Ping-Pong balls that could fit in the room you're sitting in.

First I give them about 20 seconds and ask each person to guess. Next I ask them to work in groups for 5 or 10 minutes to develop not only a numerical estimate but also a description of the method they use. At this stage, students typically model the room as a rectangular box and the ball as a cube. They then determine the number by dividing the volume of the room by the volume of a ball. I ask them what they would do if I gave them the rest of the class period to work on the problem. Sooner or later a student says, "Who cares how many Ping-Pong balls could fit in the room!" I thank that student and report that we can now stop. In any problem that involves modeling, the purpose must be specified. Without knowing the purpose, we don't know how exact an answer must be or how to use the model. In fact, the 20-second answer might be good enough.

An essential aspect of modeling is the use of heuristics (Starfield, Smith, and Bleloch, 1994), which may be generally defined as methods or strategies that aid in discovery or problem solving. Although difficult to define, heuristics are relatively easy to identify using the characteristics listed by Koen (1984, 1985, 2002):

1. Heuristics do not guarantee a solution.
2. Two heuristics may contradict each other or give different answers to the same question and still be useful.
3. Heuristics permit the solving of unsolvable problems or reduce the search time to a satisfactory solution.
4. The heuristic depends on the immediate context instead of absolute truth as a standard of validity.

Thus, a heuristic is anything that provides plausible aid or direction in the solution of a problem but is in the final analysis unjustified, incapable of justification, and fallible. It is used to guide, to discover, and to reveal. Heuristics are also a key part of the Koen's definition of the engineering method:

> The engineering method is the use of heuristics to cause the best change in a poorly understood situation within the available resources. (p. 70)

Typical engineering heuristics include (1) rules of thumb and orders of magnitude, (2) factors of safety, (3) circumstances that determine the engineer's attitude toward his or her work, (4) procedures that engineers use to keep risk within acceptable bounds, and (5) rules of thumb that are important in resource allocation.

Models and heuristics will constitute a major part of this book. The critical path method (CPM) is a procedure for modeling complex projects with interdependent activities. Visual representations include Gantt charts and network diagrams. My goal is for you to develop the skills and confidence necessary to organize, manage, be a participant in, and lead project teams. This goal is consistent with current thinking about the purpose of engineering schools. Deming associate and engineering educator Myron Tribus (1996) summarized the purpose of engineering schools as follows:

> The purpose of a School of Engineering is to teach students to create value through the design of high quality products and systems of production, and services, and to organize and lead people in the continuous improvement of these designs. (p. 25)

Notice that Tribus considers management an integral part of engineering. He also elaborates on the importance of group work for learning to engineer:

> The main tool for teaching wisdom and character is the group project. Experiences with group activities, in which the members of the groups are required to exhibit honesty, integrity, perseverance, creativity and cooperation, provide the basis for critical review by both students and teachers. Teachers will need to learn to function more as coaches and resources and less as givers of knowledge. (p. 25)

Reflection: Teamwork and Project Management in Engineering

As I finished writing this book, I was reminded of a book from 1978 that I read more than 20 years ago—*Excellence in Engineering*, by W. H. Roadstrum. I was unable to locate my copy (I probably loaned it out) but did find the second edition, *Being Successful as an Engineer* (Roadstrum, 1988). In this edition, Roadstrum remarks, "Engineering is almost completely divorced from this concept of routine and continuous. Engineering work is project work" (p. 7). Engineering *is* project work! This is the essence of Roadstrum's book. The first two chapters, "What Engineering Is" and "The Engineer," cover ground similar to the material presented in this chapter, but from a perspective about 25 years ago. Chapters 3 and 4 are "The Project and the Project Team" and ""Project Control." Although I had not looked at Roadstrum's book for several years, I was struck by the overlap between his book and mine.

Being Successful as an Engineer addresses a broad range of topics, including problem solving, laboratory work, design, research and development, manufacturing and quality control, systems, proposal work, human relations, and creativity. Roadstrum writes, "Design is the heart of the engineering process—its most characteristic activity." Furthermore, he states, "If you and I are going to understand engineering, we'll have to understand design" (p. 97).

Roadstrum further elaborates on the role of the project engineer:

> Every engineer looks forward to the time when he can have a project of his own. A project engineer has the best job in the business. He has ultimate responsibility for the work as a whole. He is the real architect of the project solution. Even more than his colleagues, he looks at the job as a whole from the beginning. He watches carefully to make all details come together into a timely, economical, fresh, and effective meeting of the need. (p. 166)

Roadstrum's book and ideas no doubt influenced my decision to develop skills and expertise in teamwork and project management; however, the specific reference lay dormant until now. I hope my book will influence your experience and practice of teamwork and project management in engineering.

A final note: This chapter opened with a discussion of Professor Billy Koen's probing questions. Koen's fourth question is this: "Lacking a ready answer [to the third question—What is the engineering method?], can you then name a nationally known engineer who is wise, well-read, and recognized as a scholar in the field of engineering—one to whom I can turn to find out what engineering really is?" To whom would you turn? Difficult, isn't it? No other profession lacks knowledgeable, clearly recognized spokespersons. I sincerely hope you'll help provide the leadership to make engineering better known.

Questions

1. What is engineering? How does engineering differ from science? What role does design play in engineering?
2. What is a model? Why are models useful in teamwork and project management and in engineering?
3. What is a system? Why are many books on teamwork and project management organized around a systems approach?

Exercises

1. Summarize your course work and experiences with engineering and design. What are some of the key things you've learned about engineers and engineering? Do you have relatives or friends who are project managers or engineers? If so, talk with them.
2. Why should you, as a first-year engineering student, be interested in teamwork and project management? Take a minute and reflect. Jot down at least three reasons why a first-year engineering student should be interested in these. What did you come up with? Did you say, for instance, that teamwork and project management are integral to professional engineering practice?
3. List your good experiences with projects and teamwork. Have you ever been on a team that had extraordinary accomplishments? If so, describe the situation, especially the characteristics of the team and project that led to extraordinary success. What were some of the factors—a sense of urgency? a project too complex or timeline too short for one person to complete? a need for synergistic interaction?

References

Accreditation Board for Engineering and Technology (ABET). 2000. *Criteria for accrediting engineering programs.* Baltimore, MD: Engineering Accreditation Commission of the Accreditation Board for Engineering and Technology.

Ackoff, Russell L. 1994. *The democratic corporation: A radical prescription for recreating corporate America and rediscovering success.* Oxford, UK: Oxford University Press.

Adams, James L. 1991. *Flying buttresses, entropy, and o-rings: The world of an engineer.* Cambridge, MA: Harvard University Press.

Bordogna, Joseph. 1998. *Realizing the New Paradigm for Engineering Education: The Professional Engineer in 2010.* Engineering Foundation Conference, Baltimore, MD, June 4, 1998.

Bucciarelli, Louis. 1994. *Designing engineers.* Cambridge, MA: MIT Press.

Burton, Lawrence, Linda Parker, and William K. LeBold. 1998. U.S. engineering career trends. *ASEE Prism* 7(9): 18–21.

Chapman, William L., A. Terry Bahill, and A. Wayne Wymore. 1992. *Engineering modeling and design.* Boca Raton, FL: CRC Press.

Churchman, C. West. 1968. *The systems approach.* New York: Laurel.

Employer's Checklist—Boeing Company. 1996. Briefings. *ASEE Prism* 6(4), 11.
Ferguson, Eugene S. 1992. *Engineering and the mind's eye.* Cambridge, MA: MIT Press.
Friedman, Thomas L. 2000. *The Lexus and the olive tree: Understanding globalization.* New York: Anchor Books.
Hancock, J. C., Chairman (1986). *Workshop on undergraduate engineering education.* Washington, DC: National Science Foundation.
Hapgood, Fred. 1992. *Up the infinite corridor: MIT and the technical imagination.* Reading, MA: Addison-Wesley.
Jordon, Brigitte. 1996. 8 principles for learning. *Fast Company,* October/November, p. 116.
Koen, B. V. 1984. Toward a definition of the engineering method. *Engineering Education* 75: 151–155.
———. 1985. *Definition of the engineering method.* Washington, DC: American Society for Engineering Education.
———. 2002. *Discussion of the method.* Oxford, UK: Oxford University Press.
Leifer, Larry. 1997. *A collaborative experience in global product-based learning.* November 18, 1997. National Technological University Faculty Forum.
Papalambros, Panos Y., and Douglass J. Wilde. 1988. *Principles of optimal design: Modeling and computation.* Cambridge, UK: Cambridge University Press.
Ray, Michael, and Alan Rinzler, eds. 1993. *The new paradigm in business: Emerging strategies for leadership and organizational change.* Los Angeles: Tarcher/Perigee.
Roadstrum, W. H. 1988. *Being successful as an engineer.* San Jose: Engineering Press.
Rothenberg, James. 1989. The nature of modeling. In *Artificial intelligence, simulation and modeling,* edited by L. E. Widman, K. A. Loparo, and N. R. Nielsen. New York: Wiley.
Senge, Peter. 1990. *The fifth discipline: The art and practice of the learning organization.* New York: Doubleday.
———. 1993. The art and practice of the learning organization. In *The New Paradigm in Business: Emerging Strategies for Leadership and Organizational Change (A New Consciousness Reader,* edited by Alan Rinzler and Michael Ray. Los Angeles: Tarcher.
Shina, S. G. 1991. New rules for world-class companies. Special Report on Concurrent Engineering, edited by A. Rosenblatt and G. F. Watson. *IEEE Spectrum* 28(7): 22–37.
Starfield, Anthony M., Karl A. Smith, and Andrew L. Bleloch. 1994. *How to model it: Problem solving for the computer age.* Edina, MN: Burgess.
Surowiecki, Jim. 1997. Decision Time. *Rogue missives.* January 6, 1997. http://www.fool.com/Rogue/1997/Rogue970106.htm (accessed 3/9/03).
Tribus, Myron. 1996. Total quality management in schools of business and engineering. In *Academic initiatives in total quality for higher education,* edited by Harry V. Roberts, 17–40. Milwaukee: ASQC Quality Press.
Weisbord, Marvin R. 1987. *Productive workplaces: Organizing and managing for dignity, meaning, and community.* San Francisco: Jossey-Bass.

CHAPTER 2

Teamwork

Everyone has to work together; if we can't get everybody working toward common goals, nothing is going to happen.

HAROLD K. SPERLICH
Former President, Chrysler Corporation

Coming together is a beginning;
Keeping together is progress;
Working together is success.

HENRY FORD

REFLECTION Think about a really effective team you've been a member of, a team that accomplished extraordinary things and perhaps was even a great place to be. Start by thinking about teams in an academic, professional, or work setting. If no examples come to mind, then think about social or community-based teams. If again you don't conjure up an example, then think about sports teams. Finally, if you don't come up with a scenario from any of these contexts, then simply imagine yourself as a member of a really effective team. OK, do you have a picture of the team in mind? As you recall (or imagine) this highly effective team experience, try to identify the specific characteristics of the team that made it so effective. Please list these characteristics.

Look over the list you made for the Reflection. Did you preface your list with "It depends"? The characteristics of an effective team depend, of course, on the purpose of the team. In large measure they depend on goals related to the team's task (what the team is to do) and maintenance (how the team functions). Michael Schrage (1991) states emphatically:

> [P]eople must understand that real value in the sciences, the arts, commerce, and, indeed one's personal and professional lives, come largely from the process of collaboration. What's more, the quality and quantity of meaningful collaboration often depend upon the tools used to create it. . . . Collaboration is a *purposive* relationship. At the heart of collaboration is a desire or need to: solve a problem, create, or discover something. (pp. 27, 34)

Let's assume that an effective team has both task goals and maintenance goals, because most effective teams have a job to do (a report to write, project to complete, presentation to give, etc.) but also a goal of getting better at working with one another.

I've used this Reflection (on preceding page) with hundreds of faculty and students in workshop and classroom settings. Here is a typical list of the characteristics of effective teams:

Good participation	Common goal
Respect	Sense of purpose
Careful listening	Good meeting facilitation
Leadership	Empowered members
Constructively managed conflict	Members take responsibility
Fun, liked to be there	Effective decision making

My goal for this chapter is to help you get a sense of the essential characteristics of teams that perform at a high level by drawing on your experience (as I have tried to do above) and start connecting you with some of the rapidly expanding literature in this area. I will remind you of the importance of maintaining a team climate that embraces and celebrates diversity. I then summarize some of the classic work on stages of team development. Finally, I aim to familiarize you with emerging notions, such as "communities of practice," "network quotient," and "emotional intelligence."

Definition of a Team

Katzenbach and Smith (1993) studied teams that performed at a variety of levels and came up with four categories:

Pseudo teams perform below the level of the average member.
Potential teams don't quite get going but struggle along at or slightly above the level of the average member.
Real teams perform quite well.
High-performing teams perform at an extraordinary level.

Katzenbach and Smith then looked for common characteristics of real teams and high-performing teams. All real teams fit this description: a small number of people with complementary skills who are committed to a common purpose, performance goals, and approach for which they hold themselves mutually accountable. High-performing teams met all the conditions of real teams and, in addition, had members who were deeply committed to one another's personal growth and success.

REFLECTION Now think about the teams in your engineering classes. Think about your most successful/effective team project experience. What were the characteristics of the team? What were the conditions? Are they similar to those of your most effective teams? Describe the team development process.

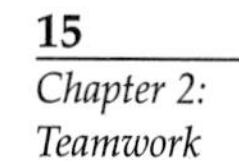

Figure 2.1 Group Performance.

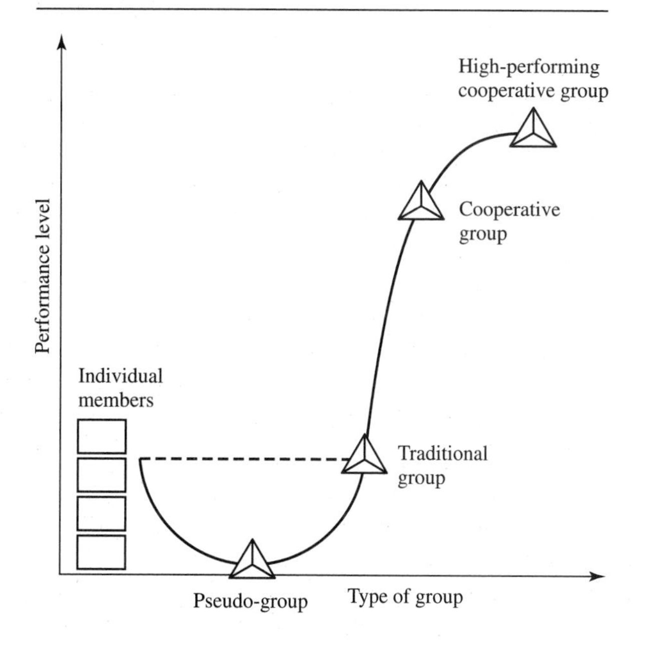

Types of Learning Teams

There is nothing magical about teamwork in engineering classes. Some types of learning teams increase the quality of classroom life and facilitate student learning. Other types of teams hinder student learning and create disharmony and dissatisfaction with classroom life. To use teamwork effectively, you must know what is and what is not an effective team.

There are many types of teams that can be used in classrooms. Formal cooperative learning groups are just one of them, although they are becoming quite common (Johnson, Johnson, and Smith, 1998b). When you choose to use (or are asked or required to use) instructional groups, you must ask yourself, What type of group or team am I involved in? Figure 2.1 and the following descriptions of groups may help you answer that question.

Pseudo Learning Group

Students in a pseudo learning group are assigned to work together but they have no interest in doing so. They believe they will be evaluated by being ranked in terms of highest performer to lowest performer. On the surface these students talk to each other, but under the surface they are competing. They see each other as rivals who must be defeated, and they block or interfere with each other's learning, hide information from each other, attempt to mislead and confuse each other, and distrust each other. These students would achieve more if they were working alone.

Traditional Classroom Learning Group

Students in a traditional classroom learning group are assigned to work together and accept that they must do so. But because assignments are structured, very little joint work is required. These students believe that they will be evaluated and rewarded as individuals, not as members of the group, so they interact primarily to clarify how assignments are to be done. They seek each other's information, but have no motivation to teach what they know to their groupmates. Helping and sharing is minimized. Some students loaf, seeking a free ride on the efforts of their more conscientious groupmates. The conscientious members feel exploited and do less. The result is that the sum of the whole is more than the potential of some of the members, but the harder working, more conscientious students would perform better if they worked alone.

Cooperative Learning Groups

Students in cooperative learning groups are assigned to work together and, given the complexity of the task and the necessity for diverse perspectives, they are relieved to do so. They know that their success depends on the efforts of all group members. The group format is clearly defined: (1) The group goal of maximizing all members' learning provides a compelling common purpose that motivates members to roll up their sleeves and accomplish something beyond their individual achievements. (2) Group members hold themselves and each other accountable for doing high-quality work to achieve their mutual goals. (3) Group members work face-to-face to produce joint work-products. They do real work together. Students promote each other's success through helping, sharing, assisting, explaining, and encouraging. They provide both academic and personal support based on a commitment to and caring about each other. (4) Group members are taught teamwork skills and are expected to use them to coordinate their efforts and achieve their goals. Both task and team-building skills are emphasized. All members share responsibility for providing leadership. (5) Groups analyze how effectively they are achieving their goals and how well members are working together. There is an emphasis on continual improvement of the quality of learning and teamwork processes. For a recent guide to success in active learning, see *Striving for Excellence in College* (Browne and Keeley, 1997).

High-Performance Cooperative Learning Group

A high-performance cooperative learning group meets all the criteria for being a cooperative learning group and outperforms all reasonable expectations, given its membership. What differentiates the high-performance group from the cooperative learning group is the level of commitment members have to each other and the group's success. Jennifer Futernick, who is part of a high-performing, rapid response team at McKinsey & Company, calls the emotional binding together of her teammates a form of love (Katzenbach and Smith, 1993). Ken Hoepner of the Burlington Northern Intermodal Transport Team stated: "Not only did we trust each other, not only did we respect each other, but we gave a damn about the rest of the people on this team. If we saw somebody vulnerable, we were there to help" (Katzenbach and Smith, 1993). Members' mutual concern for each other's personal growth enables high-performance cooperative groups to perform far above expectations, and also to have lots of fun. The bad news about extraordinarily high-performance cooperative learning groups is that they are rare. Most groups never achieve this level of development.

Groups and Teams

I've been using the term *team* in reference to projects and *group* in reference to learning, but I will use these two terms interchangeably throughout this book. Though the traditional literature focuses on groups, recently some writers have been making distinctions between groups and teams. For example, Table 2.1 presents Katzenbach and Smith's (1993) summary of the major differences between working groups and teams.

Are there any surprises in this list, from your perspective? Many students emphasize the importance of a strong leader, but Katzenbach and Smith indicate that real teams, as opposed to working groups, have shared leadership roles. Also, the literature on high-performance teams indicates that they are composed of members with complementary skills; that is, they're diverse.

Importance of Diversity

Often we must work with people who are different from us or difficult to work with but whose skills, talents, expertise, and experience are essential to the project. Working with a diverse group may seem impossible at times, but look at the example of Phil Jackson, former head coach of the Chicago Bulls basketball team. Can you imagine a more diverse group than one made up of Dennis Rodman, Michael Jordon, and Scottie Pippin? Phil Jackson is an expert at managing diversity.

Table 2.1 Not All Groups Are Teams: How to Tell the Difference

Working Group	Team
Strong, clearly focused leader	Shared leadership roles
Individual accountability	Individual and mutual accountability
The group's purpose is the same as the broader organizational mission	Specific team purpose that the team itself delivers
Individual work-products	Collective work-products
Runs efficient meetings	Encourages open-ended discussion and active problem-solving meetings
Measures its effectiveness indirectly by its influence on others	Measures performance directly by assessing collective work-products
Discusses, decides, and delegates	Discusses, decides, and does real work together

Source: Katzenbach and Smith, 1993.

Diversity has many faces, including preferred learning style (visual, auditory, kinesthetic); social background and experience; ethnic and cultural heritage; gender; and sexual orientation. The evidence from effective groups is that diversity is important—that is, the better the group represents the broader community, the more likely it is to make significant, creative, and desired contributions. Participating in and managing diverse groups is not always easy, because diverse groups usually bring a diversity of ideas and priorities. Here are some considerations that may help you learn to manage diverse groups more effectively (Cabanis, 1997; Cherbeneau, 1997):

1. Learn skills for working with all kinds of people.
2. Stress that effective teams are diverse.
3. Stress the importance of requirements.
4. Emphasize performance.
5. Develop perspective-taking skills (i.e., put yourself in others' shoes).
6. Respect and appreciate alternative perspectives.

The Chicago Bulls' former head coach Phil Jackson argued that "Good teams become great ones when the members trust each other enough to surrender the 'me' for the 'we.'" In his 1995 book (coauthored by Hugh Delehanty) *Sacred Hoops: Spiritual Lessons of a Hardwood Warrior,* Jackson offers terrific advice on organizing and managing extraordinarily high-performing teams.

Reflection: On Diversity

One of the greatest challenges I faced as co-coordinator of the Bush Faculty Development Program for Excellence and Diversity in Teaching at the University of Minnesota was helping colleagues in science, mathematics, and engineering recognize the importance of the discrepancy between the rapidly growing diversity of the population and the lack of diversity among the student body. My most memorable exposure to these issues was at a September 1997 conference held at Penn State: "Best Practices in Diversity: Exploring Practical Applications for the 21st Century." It was a real eye-opener for me to see and hear so many people deeply engaged in "making the great diversity of our nation work for the future" (in the words of Graham Spanier's welcoming remarks). The conference was particularly memorable for me because it came at the beginning of my sabbatical leave at Michigan State University, and because so many students, faculty, staff, and administrators participated in the conference that we chartered a bus and went to the conference together.

Let's start our discussion of diversity by exploring the question "Why bother?" Why should we be concerned about the diversity of student team members' experiences?

First, little attention is paid to the fact that not all students are the same. University of Minnesota astronomy professor Larry Rudnick once said, "I used to think all students learn exactly the same way I do; perhaps a little slower." It seems that many faculty assume that students are basically like themselves, not only in learning styles but in many other characteristics as well—outlook, cultural or ethnic background, experience, motivation, expectations, sexual orientation. This "sameness" approach is simpler, easier, and safer for faculty. If faculty need to design only a single, one-size-fits-all instructional system (probably the one they experienced as a student), they'll have a system that they find familiar and manageable. Faculty who acknowledge that learners are different must face lots of unknowns, and more work. But when faculty don't provide students with opportunities for diverse experiences in the classroom, students are less likely to learn the skills and knowledge they will need to work in teams with diverse membership.

The consequences of ignoring differences can be enormous. For example, they affect simple testing situations. Students from some cultures (some Native Americans and Asians, for example) are reluctant to correct others or to make them look bad in front of their peers. When there is an individual test followed by a group test format, such students might get a higher individual score but won't contradict the group during the group exam portion. Typically such a student will explain such behavior by saying that in their culture it's unacceptable to correct another person. One group dealt with this difference by always having the Asian-American students go first during the group exam portion.

Second, U.S. demographics are changing very rapidly, and undergraduate engineering enrollments don't reflect the broader diversity. Many students will choose to avoid fields of study where they don't see students like themselves enrolled, partly because they feel unwelcome.

William. A. Wulf (1998), president of the National Academy of Engineering, stressed this point in his article "Diversity in Engineering":

> Every time an engineering problem is approached with a pale, male design team, it may be difficult to find the best solution, understand the design options, or know how to evaluate the constraints. (p. 8)

Wulf also made a case for the connection between diversity and creativity:

> Collective diversity, or diversity of the group—the kind of diversity that people usually talk about—is just as essential to good engineering as individual diversity. At a fundamental level, men, women, ethnic minorities, racial minorities, and people with handicaps, experience the world differently. Those differences in experience are the "gene pool" from which creativity springs. (p. 11)

People who don't see themselves represented can find it hard to be interested in the designs, products, and services created by engineers, and engineering in turn is deprived of their marvelous talents.

Finally, diversity is the law of the land. At least three times (in *Brown v. Board of Education*, Title IX, and PL 94-142) the United States Supreme Court and Congress have reemphasized that all citizens have equal rights and opportunities—in particular, that all individuals, regardless of differences, have a right to access to the broader peer group.

Characteristics of Effective Teams

The research on highly effective teams—both in the classroom (Johnson, Johnson, and Smith, 1991, 1998a, 1998b) and in the workplace (Bennis and Biederman, 1997; Hargrove, 1998; Katzenbach and Smith, 1993; Schrage, 1991, 1995)—reveals a short list of characteristics:

1. *Positive interdependence.* The team focuses on a common goal or single product.
2. *Individual and group accountability.* Each person takes responsibility for both her or his own work and the overall work of the team.
3. *Promotive interaction.* The members do real work, usually face-to-face.
4. *Teamwork skills.* Each member has the skills for and practices effective communication (especially careful listening), decision making, problem solving, conflict management, and leadership.
5. *Group processing.* The team periodically reflects on how well the team is working, celebrates the things that are going well, and problem-solves the things that aren't.

Teams have become commonplace in engineering practice and are making inroads in engineering education. The immense literature on teams and teamwork ranges from very practical guides (e.g., Scholtes, Joiner, and Streibel, 1996; Brassand, 1995) to conceptual and theoretical treatises (e.g., Johnson and Johnson, 1991; Hackman, 1990). Check out one of these to broaden and deepen your understanding of teamwork. Four books were highlighted in this chapter—*Shared Minds: The New Technologies of Collaboration* (Shrage, 1991); *The Wisdom of Teams: Creating the High-Performance Organization* (Katzenbach and Smith, 1993); *Organizing Genius: The Secrets of Creative Collaboration* (Bennis and Biederman, 1997); and *Mastering the Art of Creative Collaboration* (Hargrove, 1998). These four books focus on extraordinary teams, teams that perform at unusually high levels and whose members experience accomplishments through synergistic interaction that they rarely experience in other settings. They provide lots of examples and insights into high-performance teams.

Building Team Performance

- Establish urgency and direction
- Select members based on skill and potential, not personalities
- Pay attention to first meeting and actions
- Set clear rules of behavior
- Set some immediate performance-oriented tasks and goals
- Challenge the group regularly with fresh information
- Spend **lots** of time together
- Exploit the power of positive feedback, recognition, and reward

Source: Katzenbach and Smith, 1994.

Stages of Team Development

Teams often progress through a series of stages. One of the most common "sequential-stage theories" was formulated by Bruce W. Tuckman (Tuckman, 1965; Tuckman and Jensen, 1977). According to Tuckman, teams develop through five sequential stages: forming, storming, norming, performing, and adjourning. Members get to know one another and start to learn to work together in the forming stage. Differences and conflicts appear during the storming stage, and much of the team's focus in the norming stage is on managing conflict. The team works together to accomplish the goals during the performing stage. The group dissolves during the adjourning stage.

An alternative to stage theory was developed by Robert Bales (1965), who argued that there must be an equilibrium between the team's focus on its task and its focus on its working relationships; that is, there must be a team maintenance orientation. Teams oscillate between a focusing on achieving their goals and focusing on maintaining good working relationships (the more emotional dimension).

Both these perspectives are valuable for understanding team development. Teams move through stages while dealing with issues that emerge. Further information on team development is available in *The Team Developer* (McGourty and DeMeuse, 2001), *Joining Together* (Johnson and Johnson, 1991), and *The Team Handbook* (Scholtes, Joiner, and Streibel, 1996).

Emerging Ideas

Other exciting developments in the area of teamwork include the emerging ideas of communities of practice, emotional intelligence, and network quotient. Communities of practice are essential in many companies (e.g., Boeing, Daimler Chrysler) for managing and developing knowledge. Here's a definition of such communities:

> Communities of practice are groups of people who share a concern, a set of problems, or a passion about a topic, and who deepen their knowledge and expertise in this area by interacting on an ongoing basis. (Wenger, McDermott, and Snyder, 2002, p. 4)

The concept of emotional intelligence is also being heralded as important for team and project success. Daniel Goleman (1998) defines emotional intelligence as "the capacity for recognizing our own feelings and those of others, for motivating ourselves, and for managing emotions well in ourselves and in our relationships" (p. 24).

A related idea is that of NQ, or network quotient. Tom Boyle of British Telecom, who calls this the age of interdependence, says that people's NQ—their capacity to form connections with one another—is now more important than IQ, the measure of individual intelligence (Cohen and Prusak, 2001).

These emerging ideas indicate that teamwork, project management, and knowledge management are dynamic areas where there is a lot of innovation. So, stay posted and stay alert.

Effective teamwork is not easy to accomplish. Engineering professor Douglas J. Wilde said, "It's the soft stuff that's hard, the hard stuff is easy." (Leifer, 1997) Larry Leifer (1997), director of the Stanford Center for Design Research, reports, "Design team failure is usually due to failed team dynamics." However, if you work at it, continue to study and learn about effective teamwork, and attend to the skills and strategies needed for effective teamwork described in Chapter 3, you will very likely have many positive team experiences.

Reflection: Interdependence and Teamwork

I've been a student of interdependence and teamwork ever since I took a course on the social psychology of education in about 1974. Prior to that I had predominantly thought of learning (and work, for that matter) as an individual endeavor. The instructor of that course, Dennis Falk, one of David Johnson's graduate students, had us working together, cooperatively; and he emphasized positive interdependence. I had an epiphany! I thought, This is the way I worked as an engineer—why isn't the classroom organized in this way? Numerous resources are available to help faculty organize and manage learning teams. Especially see those developed by the Foundation Coalition Active/Cooperative Learning Project (www.foundationcoalition.org), which are intended to help students learn how to work together.

I've often wondered why there was such an emphasis on interdependence in Minnesota. I haven't discovered the answer yet, but it might be due to the Lakota presence in Minnesota. One of the cornerstones of Lakota culture is the phrase used in all their ceremonies—*mitakuye oyasin* ("We are all related") (Marshall, 2001). According to Medearis and White Hat (1995), the connection between *mitakuye oyasin* and education is this: "Education is an art of process, participation, and making connection. Learning is a growth and life process; and life and Nature are always relationships in process" (p. 1).

Questions

1. What are the characteristics of effective teams? How do you help promote them?
2. Where and how have teamwork skills been taught or emphasized to you? in school? in social groups? in professional groups? in your family? Describe two or three instances where teamwork skills were emphasized.
3. How is increasing ethnic diversity affecting project teams? What are some strategies for effectively participating on and managing diverse teams?
4. Students often say that groups in school are different from groups in the workplace, giving this as a reason for not using groups in school. Is it a valid excuse? Summarize the major differences between groups in school and groups in the workplace. How are these differences beneficial or harmful to the work of the group? What are some things you can do to improve the school groups?

Exercises

1. Check out a study of teams that have performed at extraordinary levels. Some of the books listed in the references for this chapter have terrific stories of stellar teams (see, e.g., Hargrove, 1998; Bennis and Biederman, 1997; Schrage, 1991, 1995). You may want to check the library or do an electronic search of the literature. Summarize the features of these extraordinary teams. How does your summary compare with the list provided in this chapter? Remember, this is a dynamic area of research with lots of new books and articles appearing each year.
2. Look for opportunities to participate on a superb team. Make a plan for participating on a high-performance team.
3. Study the diversity of teams in your school or workplace and note strategies for recognizing, valuing, and celebrating diversity.

References

Bales, Robert F. 1965. The equilibrium problem in small groups. In *Small groups: Studies in social interaction,* edited by A. Hare, E. Borgatta, and R. Bales. New York: Knopf.

Bennis, Warren, and Patricia Biederman. 1997. *Organizing genius: The secrets of the creative collaboration.* Reading, MA: Addison-Wesley.

Brassand, Michael. 1995. *The team memory jogger: A pocket guide for team members.* Madison, WI: GOAL/QPC and Joiner Associates.

Browne, M. Neil, and Stuart Keeley. 1997. *Striving for excellence in college.* Upper Saddle River, NJ: Prentice Hall.

Cabanis, Jeannette. 1997. Diversity: This means you. *PM Network* 11(10): 29–33.

Cherbeneau, Jeanne. 1997. Hearing every voice: How to maximize the value of diversity on project teams. *PM Network* 11(10): 34–36.

Cohen, Don, and Laurence Prusak. 2001. *In good company: How social capital makes organizations work.* Cambridge, MA: Harvard Business School Press.

Covey, Stephen R. 1989. *The seven habits of highly effective people.* New York: Simon & Schuster.

Goleman, Daniel. 1998. *Working with emotional intelligence.* New York: Bantam Books.

Hackman, J. R. 1990. *Groups that work (and those that don't): Creating conditions for effective teamwork.* San Francisco: Jossey-Bass.

Hargrove, Robert. 1998. *Mastering the art of creative collaboration.* New York: McGraw-Hill.

Jackson, Phil, and Hugh Delehanty. 1995. *Sacred hoops: Spiritual lessons of a hardwood warrior.* New York: Hyperion.

Johnson, David W., and Frank P. Johnson. 1991. *Joining together: Group theory and group skills,* 4th ed. Englewood Cliffs, NJ: Prentice Hall.

Johnson, David W., Roger T. Johnson, and Karl A. Smith. 1991. *Cooperative learning: Increasing college faculty instructional productivity.* Washington, DC: ASHE-ERIC Reports on Higher Education.

———. 1998a. *Active learning: Cooperation in the college classroom,* 2nd ed. Edina, MN: Interaction Book.

———. 1998b. Maximizing instruction through cooperative learning. *ASEE Prism* 7(6): 24–29.

Katzenbach, Jon, and Douglas Smith. 1993. *The wisdom of teams: Creating the high-performance organization.* Cambridge, MA: Harvard Business School Press.

Marshall, Joseph M. III. 2001. *The Lakota way: Stories and lessions for living.* New York: Penguin.

McGourty, Jack, and Kenneth P. DeMeuse. 2001. *The team developer: An assessment and skill building program.* New York: Wiley.

Medearis, Cheryl, and Albert White Hat, Sr. 1995. *Mitakuye oyasin.* Paper presented at the conference. Collaboration for the Advancement of College Teaching and Learning Faculty Development, Minneapolis, MN, November, 1995.

Scholtes, Peter R., Brian L. Joiner, and Barbara J. Streibel. 1996. *The team handbook.* Madison, WI: Joiner Associates.

Schrage, Michael. 1991. *Shared minds.* New York: Random House.

———. 1995. *No more teams! Mastering the dynamics of creative collaboration.* New York: Doubleday.

Tuckman, Bruce. 1965. Development sequence in small groups. *Psychological Bulletin* 63: 384–399.

Tuckman, Bruce, and M. Jensen. 1977. Stages of small group development revisited. *Groups and Organizational Studies* 2: 419–427.

Wenger, Etienne, Richard McDermott, and William Snyder. 2002. *Cultivating communities of practice.* Cambridge, MA: Harvard Business School Press.

Wulf, William A. 1998. Diversity in engineering. *Bridge* 28(4): 8–13.

CHAPTER 3

Teamwork Skills and Problem Solving

I will pay more for the ability to deal with people than any other ability under the sun.

JOHN D. ROCKEFELLER

If you can't operate as a team player, no matter how valuable you've been, you really don't belong at GE. (1993)

JOHN F. WELCH
CEO, General Electric

REFLECTION Have you been a member of a team that got the job done (wrote the report, finished the project, completed the laboratory assignment) but that ended up with the members hating one another so intensely they never wanted to see each other again? Most students have, and they have found it very frustrating. Similarly, have you been a member of a team whose members really enjoyed one another's company and had a great time socially, but in the end hadn't finished the project? Again, most students have, and they also have found this frustrating. Take a moment to recall your experiences with these two extremes of teamwork.

Importance of Task and Relationship

As noted in Chapter 2, to be most effective, groups need to do two things very well: accomplish the task and get better at working with one another. Both of these require leadership—not just from a single person acting as the leader but from every member contributing to the leadership of the group. This chapter focuses on teamwork skills using a "distributed actions approach" to leadership. *Distributed actions* are specific behaviors that group members engage in to help the group accomplish its task or to improve working relationships. Napier and Gershenfeld (1973) summarize many of these behaviors (see Table 3.1). Note the date—1973—which indicates that effective group work is not a new concept.

Table 3.1 Group Task and Maintenance Roles

Group Task Roles	Group Maintenance Roles
Initiating	Encouraging
Seeking information	Expressing feelings
Giving information	Harmonizing
Seeking opinions	Compromising
Giving opinions	Facilitating communications
Clarifying	Setting standards or goals
Elaborating	Testing agreement
Summarizing	Following

Table 3.2 Management Behavior Change Needed for Team Culture

From	To
Directing	Guiding
Competing	Collaborating
Relying on rules	Focus on the process
Using organizational hierarchy	Using a network
Consistency/sameness	Diversity/flexibility
Secrecy	Openness/sharing
Passive	Risk taking
Isolated decisions	Involvement of others
People as costs	People as assets
Results thinking	Process thinking

Source: McNeill, Bellamy, and Foster, 1995.

To achieve the benefits of a team culture, some changes in management behavior are needed, as shown in Table 3.2. To learn more about the behaviors listed on the right-hand side of Table 3.2, read on.

Organization—Group Norms

A common way to promote more constructive and productive teamwork is to have the team create a set of guidelines for the group, sometimes called group norms. Take a minute and list some things (attitudes, behaviors, and so on) that you have found (or believe) can help a group be more effective. Then compare your list with the following two lists, both of which are from McNeill, Bellamy, and Foster (1995). The first was adapted from the Boeing Airplane Group's training manual for team members and the second is from the Ford Motor Company.

Code of Cooperation

1. *Every* member is responsible for the team's progress and success.
2. Attend all team meetings and be on time.
3. Come prepared.
4. Carry out assignments on schedule.
5. Listen to and show respect for the contributions of other members; be an active listener.

6. *Constructively* criticize ideas, not persons.
7. Resolve conflicts constructively.
8. Pay attention; avoid disruptive behavior.
9. Avoid disruptive side conversations.
10. Only one person speaks at a time.
11. Everyone participates; no one dominates.
12. Be succinct; avoid long anecdotes and examples.
13. No rank in the room.
14. Respect those not present.
15. Ask questions when you do not understand.
16. Attend to your personal comfort needs at any time, but minimize team disruption.
17. Have fun.
18. ?

Ten Commandments: An Affective Code of Cooperation

- Help each other be right, not wrong.
- Look for ways to make new ideas work, not for reasons they won't.
- If in doubt, check it out. Don't make negative assumptions about each other.
- Help each other win, and take pride in each other's victories.
- Speak positively about each other and about your organization at every opportunity.
- Maintain a positive mental attitude no matter what the circumstances.
- Act with initiative and courage, as if it all depends on you.
- Do everything with enthusiasm; it's contagious.
- Whatever you want, give it away.
- Don't lose faith.
- Have fun!

Team norms are common today not only in business and industry, but also in academic and research settings. The box "Tips for Working Successfully in a Group" presents a list developed by Randy Pausch for use in a course he taught at Carnegie Mellon University (Pausch, 2002). (Pausch's website also contains a set of slides that summarize his terrific, and radical, ideas on time and project management—see http://wonderland.hcii.cs.cmu.edu/randy.htm.) Having an agreed-upon, abided-by code of cooperation such as Pausch's will help groups get started toward working effectively. However, if group members haven't developed the requisite communication, trust, loyalty, organization, leadership, decision-making procedures, and conflict management skills, then the group will very likely struggle or at least not perform up to its potential. One way a team can develop such a code is to create a *team charter*—a sample format for a team charter is given below. Also see Exercise 3 at the end of this chapter.

Team Charter Guidelines

- Team name, membership, and roles
- Team mission statement
- Anticipated results (goals)
- Specific tactical objectives
- Ground rules/guiding principles for team participation
- Shared expectations/aspirations

Team charters typically are created during a team meeting early in the project life cycle. Involvement of all team members in creating the charter helps build commitment of each to the project and to other members. A set of guidelines such as the Team Charter Guidelines often helps the team through this process.

Let's start to look more deeply into the mystery of teamwork skills, starting with a summary of our work on teamwork skills in learning groups.

Tips for Working Successfully in a Group

Meet people properly. It all starts with the introduction. Then exchange contact information and make sure you know how to pronounce everyone's names. Exchange phone numbers, and find out when it is acceptable to call.

Find things you have in common. You can almost always find something in common with another person, and starting from that baseline, it's much easier to then address issues where you have differences. This is why cities like professional sports teams, which are socially galvanizing forces that cut across boundaries of race and wealth. If nothing else, you probably have in common things like the weather.

Make meeting conditions good. Have a large surface to write on, make sure the room is quiet and warm enough, and that there aren't lots of distractions. Make sure no one is hungry, cold, or tired. Meet over a meal if you can; food *softens* a meeting. That's why they "do lunch" in Hollywood.

Let everyone talk. Even if you think what they're saying is stupid. Cutting someone off is rude, and not worth whatever small time gain you might make. Don't finish someone's sentences for him or her; they can do it for themselves. And remember: talking louder or faster doesn't make your idea any better.

Check your egos at the door. When you discuss ideas, immediately label them and write them down. The labels should be descriptive of the idea, not the originator: "the troll bridge story," not "Jane's story."

Praise each other. Find something nice to say, even if it's a stretch. Even the worst of ideas has a silver lining inside it, if you just look hard enough. Focus on the good, praise it, and then raise any objections or concerns you have about the rest of it.

Put it in writing. Always write down who is responsible for what, by when. Be *concrete.* Arrange meetings by e-mail, and establish accountability. Never assume that someone's roommate will deliver a phone message. Also, remember that "politics is when you have more than 2 people"—with that in mind, always copy any piece of e-mail within the group, or to me, to *all members* of the group. This rule should *never* be violated; don't try to guess what your group mates might or might not want to hear about.

Be open and honest. Talk with your group members if there's a problem, and talk with me if you think you need help. The whole point of this course is that it's tough to work across cultures. If we all go into it knowing that's an issue, we should be comfortable discussing problems when they arise—after all, that's what this course is really about. Be forgiving when people make mistakes, but don't be afraid to raise the issues when they come up.

Avoid conflict at all costs. When stress occurs and tempers flare, take a short break. Clear your heads, apologize, and take another stab at it. Apologize for upsetting your peers, even if you think someone else was primarily at fault; the goal is to work together, not start a legal battle over whose transgressions were worse. It takes two to have an argument, so be the peacemaker.

Phrase alternatives as questions. Instead of "I think we should do A, not B," try "What if we did A, instead of B?" That allows people to offer comments, rather than defend one choice.

Source: Randy Pausch, for the *Building Virtual Worlds* course at Carnegie Mellon, Spring 1998.

Teamwork Skills

What are teamwork skills, and how does one learn them? This is an area we've researched in our study of active and cooperative learning (Johnson, Johnson, and Smith, 1998). We identified the following categories of skills—forming, functioning, formulating, and fermenting—and have suggestions for mastering them.

Cooperative Teamwork Skills

Forming skills—*Initial management skills*

- Move into groups quietly.
- Stay with the group.
- Use quiet voices.
- Take turns.
- Use names, look at the speaker.
- No "put-downs."

Functioning skills—*Group management skills*

- Share ideas and opinions.
- Ask for facts and reasoning.
- Give direction to the group's work (state assignment purpose, provide time limits, offer procedures).
- Encourage everyone to participate.
- Ask for help or clarification.
- Express support and acceptance.
- Offer to explain or clarify.
- Paraphrase others' contributions.
- Energize the group.
- Describe feelings when appropriate.

Formulating skills—*Formal methods for processing materials*

- Summarize out loud completely.
- Seek accuracy by correcting/adding to summaries.
- Help the group find clever ways to remember.
- Check understanding by demanding vocalization.
- Ask others to plan for telling/teaching out loud.

Fermenting skills—*Stimulating cognitive conflict and reasoning*

- Criticize ideas without criticizing people.
- Differentiate the ideas and the reasoning of members.
- Integrate ideas into single positions.
- Ask for justification of conclusions.
- Extend answers.
- Probe by asking in-depth questions.
- Generate further answers.
- Test reality by checking the group's work.

Learning Cooperative Teamwork Skills

1. Observe and reflect to see the *need* to learn the skill.
2. *Learn how* to do it (T-chart—what does it look like, what does it sound like).
3. *Practice* the skill daily.
4. *Reflect* on, process, and refine use.
5. *Persevere* until skill is automatic.

These cooperative teamwork skills are essential for productive and successful teamwork, and they must be learned and practiced with the same seriousness with which other engineering skills are learned.

Communication

Effective communication—listening, presenting, persuading—is at the heart of effective teamwork. The task and maintenance roles we have listed all involve oral communication. Here are the listening skills emphasized in Arizona State University's course Introduction to Engineering Design (McNeill, Bellamy, and Foster, 1995):

Stop talking.
Engage in one conversation at a time.
Empathize with the person speaking.
Ask questions.
Don't interrupt.
Show interest.
Concentrate on what is being said.
Don't jump to conclusions.
Control your anger.
React to ideas, not to the speaker.
Listen for what is not said; ask questions.
Share the responsibility for communication.

Three listening techniques they recommend are these:

Critical listening
- Separate fact from opinion.

Sympathetic listening
- Don't talk—listen.
- Don't give advice—listen.
- Don't judge—listen.

Creative listening
- Exercise an open mind.
- Supplement your ideas with another person's ideas and vice versa.

You may be wondering why so much emphasis on listening. The typical professional spends about half of his or her business hours listening, and project managers may spend an even higher proportion of their time listening. Most people, however, are not 100 percent efficient in their listening. Typical listening efficiencies are only 25 percent (Taylor, 1998). The first list provides suggestions to help the listener truly hear what is being said, and the second highlights the fact that different situations call for different types of listening.

REFLECTION Take a moment to think about the listening skills and techniques. Do you listen in all three ways listed above? Which are you best at? Which do you need to work on?

Leadership

A common notion is that leadership is a trait that some are born with. Another common notion is that a person's leadership ability depends on the situation. There is an enormous literature on leadership, so I'll only provide insights that I've found useful. I'll also try to guide you to more reading and resources on the topic.

> INDIVIDUAL AND GROUP REFLECTION What does it mean to lead a team? What does it take? Take a moment to reflect on the characteristics you admire most in a leader. Jot down 8 to 10 of them. Compare your list with your teammates' lists.

Leadership authors Kouzes and Posner (1987, 1993) have asked thousand of people to list the characteristics of leaders they admire. Table 3.3 lists the most common responses from their 1987 and 1993 studies. Many students and workshop participants express surprise that honesty is listed as number one. They say it's a given. Apparently honesty is not a given for many leaders in business and industry. In 1993, Kouzes and Posner also asked the respondents to list the most desirable characteristics of colleagues. Honest was number one again, with 82 percent selecting it. Cooperative, dependable, and competent were second, third, and fourth, with slightly more than 70 percent of respondents selecting each.

Table 3.3 Characteristics of Admired Leaders

Characteristic	1987 U.S. Percentage of People Selecting	1993 U.S. Percentage of People Selecting
Honest	83	87
Forward-looking	62	71
Inspiring	58	68
Competent	67	58
Fair-minded	40	49
Supportive	32	46
Broad-minded	37	41
Intelligent	43	38
Straightforward	34	34
Courageous	27	33
Dependable	32	32
Cooperative	25	30
Imaginative	34	28
Caring	26	27
Mature	23	14
Determined	20	13
Ambitious	21	10
Loyal	21	10
Self-controlled	13	5
Independent	13	5

Source: Kouzes and Posner, 1987, 1993.

Kouzes and Posner found that when leaders do their best, they challenge, inspire, enable, model, and encourage. They suggest five practices and ten behavioral commitments of leadership:

Challenging the Process

1. Search for opportunities.
2. Experiment and take risks.

Inspiring a Shared Vision

3. Envision the future.
4. Enlist others.

Enabling Others to Act

5. Foster collaboration.
6. Strengthen others.

Modeling the Way

7. Set the example.
8. Plan small wins.

Encouraging the Heart

9. Recognize individual contributions.
10. Celebrate accomplishments.

Peter Scholtes, author of the best-selling book *The Team Handbook,* recently published *The Leader's Handbook* (Scholtes, 1998). He offers the following six "New Competencies" for leaders:

1. The ability to think in terms of systems and knowing how to lead systems.
2. The ability to understand the variability of work in planning and problem solving.
3. Understanding how we learn, develop, and improve; leading true learning and improvement.
4. Understanding people and why they behave as they do.
5. Understanding the interaction and interdependence between systems, variability, learning, and human behavior; knowing how each affects the others.
6. Giving vision, meaning, direction, and focus to the organization.

☞ REFLECTION Take a moment to reflect on what you've learned thus far about the competencies Scholtes emphasizes—systems, thinking, variability, learning and improvement, understanding people, interdependence, and giving vision—and list connections both with your personal experiences and with earlier sections of this book.

The latest breakthrough work on leadership is Jim Collins's concept of Level 5 leadership (Collins, 2001a, 2001b). Collins and his research team studied companies who moved from being good to being great. Their central finding was that the leaders of these companies "build enduring greatness through a paradoxical combination of personal humility plus professional will."

Collins's revelation reminds me of a virtue that philosopher Walter Kaufmann said is a cardinal virtue—the fusion of humility and ambition (Kaufmann, 1973). I've tried to live by Kaufmann's cardinal virtue of fusing humility and ambition for the past 25 years. I find it interesting that Collins's work has focused on a similar fusion. There is something significant here, and I suggest that you reflect on it.

In addition to group norms, communication, and leadership, teamwork depends on effective decision making and constructive conflict management, described in the next two sections.

Decision Making

There are several approaches to making decisions in groups. Before exploring them, however, I suggest that you try a group decision-making exercise. Common exercises to assist in the development of teamwork skills—especially communication (sharing knowledge and expertise), leadership, and decision making—are ranking tasks, such as survival tasks, in which a group must decide which items are most important for survival in the desert, on the moon, or in some other difficult place. Ranking tasks are common in organizations that must select among alternative designs, hire personnel, or choose projects or proposals for funding.

My favorite ranking task for helping groups focus on communication, leadership, decision making, and conflict resolution is "They'll Never Take Us Alive." This exercise, which includes both individual and group decision making, is Exercise 1 at the end of this chapter (see page 41). Do that exercise now.

GROUP REFLECTION 1: How did your group make the decision? Did you average your individual rankings? vote? Did you discuss your individual high and low rankings and then work from both ends toward the middle? Did you try to reach consensus? Were you convinced by group members who seem to have "expert" knowledge? Did you start with the number of fatalities for one of the activities and work from there?

GROUP REFLECTION 2: How well did your group work? What went well? What things could you do even better next time?

What method a group uses to make a decision depends on many factors, including how important the decision is and how much time there is to decide. Groups should have a good repertoire of decision-making strategies and a means of choosing the one that is most appropriate for the situation.

Several methods have been described in the literature for making decisions. One of my favorites is from David Johnson and Frank Johnson (1991). The authors list seven methods for making decisions:

1. *Decision by authority without discussion.* The leader makes all the decisions without consulting the group. This method is efficient but does not build team member commitment to the decision.
2. *Expert member.* The most expert member is allowed to decide for the group. The difficulty with this method often lies in deciding who has the most expertise, especially when those with power or status in the group overestimate their expertise.
3. *Average of members' opinions.* The group decision is based on the average of individual group members' opinions.
4. *Decision by authority after discussion.* The designated leader makes the decision after discussion with the group. The effectiveness of this method often depends on the listening skills of the leader.
5. *Minority control.* Two or more members who constitute less than 50 percent of the group often make decisions by acting as (a) an executive committee or (b) a special problem-solving subgroup.
6. *Majority control.* Decision by a majority vote is the method used by the U.S. Congress. Discussion occurs only until at least 51 percent of the members agree on a course of action.
7. *Consensus.* Consensus is probably the most effective method of group decision making in terms of decision quality and gaining members' commitment to the decision, but it also may take the most time. Perfect consensus is achieved when everyone agrees. A lesser degree of consensus is often accepted where everyone has had their say and will commit to the decision, even though not everyone completely agrees with the decision.

David and Frank Johnson (1991) note that the quality of the decision and the time needed vary as a function of the level of involvement of the people involved in the decision-making method, as shown in Figure 3.1.

They also list the following characteristics of effective decisions:

1. The resources of the group members are well used.
2. Time is well used.
3. The decision is correct, or of high quality.
4. The decision is put into effect fully by all the necessary members' commitments.
5. The problem-solving ability of the group is enhanced.

> GROUP REFLECTION How well did your group do on each of these five characteristics of effective decisions?

Typically, novice decision-making groups don't take full advantage of the skills and talents of their members, and they often struggle to get started. Some researchers report a series of stages in team development (e.g., forming, storming, norming, performing) and offer suggestions for working through each stage (Scholtes, Joiner, and Streibel, 1996). Also, if you ask a group to invest time and effort in making a decision, it is very important that the decision or recommendation of the group be implemented (or that very good rationale be

Figure 3.1 Decision Type and Quality.

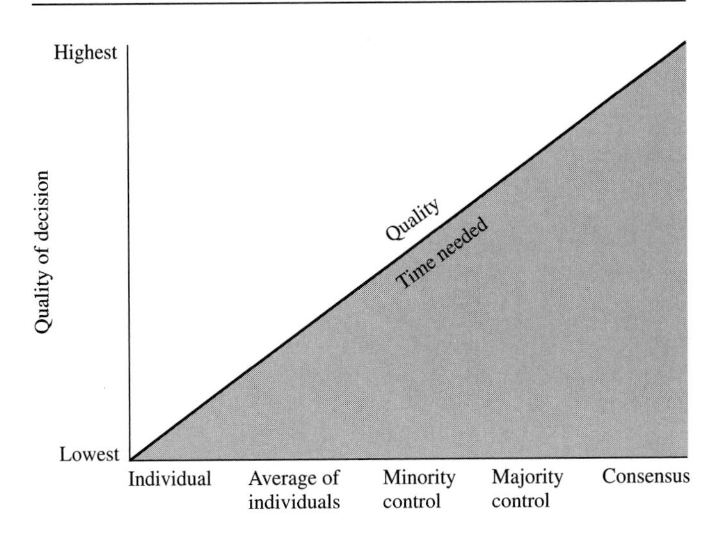

provided for why it wasn't implemented). Few things are more frustrating than to be asked to spend a lot of time and effort on work that goes nowhere.

Some of the latest and most interesting work on decision making comes from David Garvin and Michael Roberto (2001). They propose that we view decision making as an inquiry process rather than as an advocacy process, so that decision making is seen as collaborative problem solving rather than as a contest. Key differences between an advocacy approach and an inquiry approach are shown in Table 3.4. Their inquiry approach to decision making is consistent with a constructive academic controversy approach my colleagues and I devised to help students learn about reaching decisions in controversial situations (Johnson, Johnson, and Smith, 2000). Approaching, framing, and working through decisions in this manner can result in far more enlightened and constructive decisions.

Table 3.4 Two Approaches to Decision Making

	Advocacy	**Inquiry**
Concept of decision making	A contest	Collaborative problem solving
Purpose of discussion	Persuasion and lobbying	Testing and evaluation
Participants' role	Spokespeople	Critical thinkers
Pattern of behavior	Strive to persuade others Defend your position Downplay weaknesses	Present balanced arguments Remain open to alternatives Accept constructive criticism
Minority views	Discouraged or dismissed	Cultivated and valued
Outcome	Winners and losers	Collective ownership

Source: Garvin and Roberto, 2001.

Russo and Shoemaker (2002) describe an interesting and straightforward four-step decision-making process:

1. *Frame.* Decide what you are going to decide and what you are not going to decide.
2. *Gather intelligence.* Gather real intelligence, not just information that will support your biases.
3. *Come to conclusions.* Determine how your group will act on the intelligence it gathers.
4. *Learn from experience.*

Russo and Shoemaker's approach helps demystify the process of decision making. Their guidance through each of the steps provides insight into the process and highlights key concepts. They also provide case studies and worksheets to help readers apply the approach to their own decision-making situations.

Making decisions and providing information so that others can make decisions are two of the most important and common activities of practicing engineers.

Conflict Management

Conflict is a routine aspect of every project manager's job. *Conflict* is a situation in which an action of one person prevents, obstructs, or interferes with the actions of another person. On complex projects and tasks, highly talented and motivated people routinely disagree about the best ways to accomplish tasks and especially about how to deal with trade-offs among priorities. A conflict often is a moment of truth, because its resolution can follow either a constructive or a destructive path.

> *The work life of a project manager is a life of conflict. Although conflict is not necessarily bad, it is an issue that has to be resolved by the project manager. Without excellent negotiation skills, the project manager has little chance for success.*
>
> JAMES TAYLOR
> A Survival Guide for Project Managers

INDIVIDUAL REFLECTION Write the word *conflict* in the center of a blank piece of paper and draw a circle around it. Quickly jot down all the words and phrases you associate with the word *conflict* by arranging them around your circle.

Review your list of associations and categorize them as positive, negative, or neutral. Count the total number of positive, negative, and neutral associations, and calculate the percentage that are positive. Did you have more than 90 percent positive?

Fewer than 5 percent of the people I've worked with in classes and workshops have had higher than 90 percent positive associations with the word *con-*

flict. Most, in fact, have had lower than 50 percent positive associations. Many have lower than 10 percent positive.

The predominance of negative associations with conflict is one of the reasons conflict management is so difficult for project managers. Many people prefer to avoid conflict or suppress it when it does arise. They become fearful, anxious, angry, or frustrated; consequently, the conflict takes a destructive path.

The goal of this section is to help you develop a set of skills and procedures for guiding conflict along a more constructive path. I'd like to begin by asking you to complete a questionnaire to assess how you typically act in conflict situations. The "How I Act in Conflict" questionnaire is included as Exercise 2 at the end of this chapter (page 42). Take a few minutes to complete and score the questionnaire. Try to use professional conflicts and not personal conflicts as your point of reference.

Set the questionnaire aside for a few minutes and read Exercise 3, the Ralph Springer case study (page 43). Work through the exercise, completing the ranking form at the end.

GROUP ACTIVITY Share and discuss each member's results from Exercise 2. Discuss each of the possible ways to resolve the conflict.

Then compare your individual responses from Exercise 2 to your rankings in Exercise 3. Note that each of the alternatives listed in Exercise 3 represents one of the strategies listed on the scoring form in Exercise 2. Match the alternatives to the strategies they represent. Discuss similarities and differences in the order in which each group member would have used the strategies and the relative effectiveness of each.

The five conflict strategies shown in Exercise 2—withdrawal, forcing, smoothing, compromise, and confrontation—were formulated into a model for analyzing approaches to conflict by Blake and Mouton (1964). The authors used two axes to represent the conflict strategies: (1) The importance of the goal, and (2) the importance of the relationship. The placement of each of the five strategies according to this framework is shown in Figure 3.2. The five conflict strategies are described as follows:

1. *Withdrawal.* Neither the goal nor the relationship is important—you withdraw from the interaction.
2. *Forcing.* The goal is important but not the relationship—use all your energy to get the task done.
3. *Smoothing.* The relationship is more important than the goal. You want to be liked and accepted.
4. *Compromise.* Both goal and relationship are important, but there is a lack of time—you *both* gain and lose something.
5. *Confrontation.* Goal and relationship are equally important. You define the conflict as a problem-solving situation and resolve through negotiation.

Each of these strategies is appropriate under certain conditions. For example, if neither the goal nor the relationship is important to you, then often the

Figure 3.2 Blake and Mouton Conflict Model.

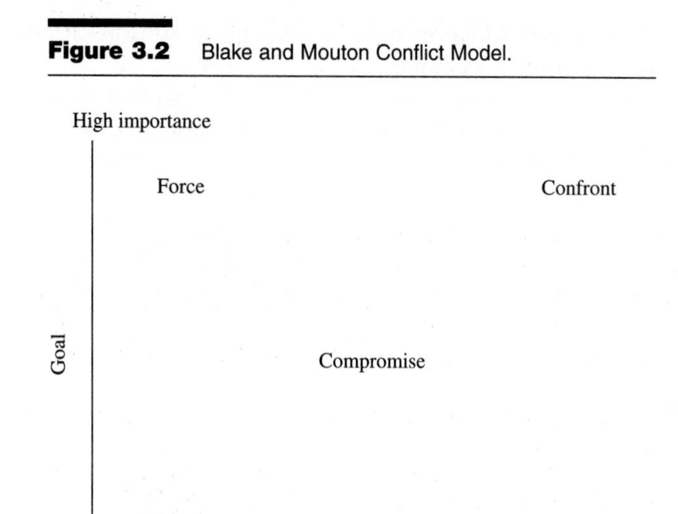

best thing to do is withdraw. If the relationship is extremely important and the goal is not so important (at the time), then smoothing is appropriate. In many conflict situations, both the goal and the relationship are important. In these situations, the strategy of confronting and negotiating often leads to the best outcomes.

A *confrontation* is the direct expression of one opponent's view of the conflict, and his or her feelings about it, and an invitation to the other opponent to express their view and feelings.

Guidelines for Confrontation

1. Do not "hit-and-run." Confront only when there is time to jointly define the conflict and schedule a negotiating session.
2. Openly communicate your feelings about and perceptions of the issues involved in the conflict, and try to do so in minimally threatening ways.
3. Accurately and fully comprehend the opponent's views and feelings about the conflict.

Negotiation is a conflict resolution process by which people who want to come to an agreement try to work out a settlement.

Steps in Negotiating a Conflict

1. Confront the opposition.
2. Define the conflict mutually.
3. Communicate feelings and positions.
4. Communicate cooperative intentions.
5. Take the other person's perspective.
6. Coordinate the motivation to negotiate.
7. Reach an agreement that is satisfactory to both sides.

Constructively resolving conflicts through a confrontation–negotiation process takes time and practice to perfect, but it's worth it. Conflicts that do not get resolved at a personal level must be resolved at more time-consuming and costly levels—third-party mediation, arbitration, and, if all else fails, litigation.

Finally, here are some heuristics for dealing with conflicts in long-term personal and professional relationships:

1. Do not withdraw from or ignore the conflict.
2. Do not engage in "win–lose" negotiations.
3. Assess for smoothing.
4. Compromise when time is short.
5. Confront to begin problem-solving negotiations.
6. Use your sense of humor.

Remember that heuristics are reasonable and plausible, but not guaranteed. I suggest that you develop your own set of heuristics for dealing with conflict as well as for the other skills needed for effective teamwork. Some of my former students who now work as project managers emphasize during classroom visits that they spend a lot of time resolving conflicts—over meeting specifications, schedules, delivery dates, interpersonal problems among team members—and that they deal with most conflicts informally.

Teamwork Challenges and Problems

> REFLECTION What are some of the most common challenges and problems you've had working in teams? Please reflect for a moment. Make a list. Has a professor ever had you do this in your teams? If so, it's a clear indication that the professor understands the importance of group processing for identifying and solving problems.
>
> What's on your list?

The challenges and problems you listed in the above Reflection may have included the following:

- Members who don't show up for meetings or who show up unprepared.
- Members who dominate the conversation.
- Members who don't participate in the conversation.
- Time wasted by off-task talk.
- Members who want to do the entire project themselves because they don't trust others.
- Group meeting scheduling difficulties.
- No clear focus or goal.
- Lack of clear agenda, or hidden agendas.
- Subgroups excluding or ganging up on one or more members.
- Ineffective or inappropriate decisions and decision-making processes.
- Suppression of conflict or unpleasant flareups among group members.
- Members not doing their fair share of the work.
- Lack of commitment to the group's work by some members.

These problems are commonly encountered by students (and professionals) working in teams and groups. If they are not addressed they can turn a cooperative group into a pseudo group (as described in Chapter 2), where the group performs worse than the individuals would have performed if working alone. If the challenges are addressed in a problem-solving manner, then the group is likely to perform at much higher levels (and the members will have a much more positive experience). The following process is widely used to address group problems.

Step 1: Identifying Challenges, Difficulties, and Barriers to Effective Group Work: Develop a List of Challenges, Barriers, and Problems

- Reflect individually for a moment and start a list of challenges, barriers, or problems facing the group. Share the individual lists and create a joint list that includes at least one item from each group member.
- Do not solve (yet).
- Be realistic and specific.
- Work cooperatively.
- If more than one group is involved, list challenges, barriers, and problems for all groups on an overhead projector or flip chart.

Step 2: Addressing Barriers, Challenges, and Problems

1. Have each group or (if only one group is involved) each member select one item from the joint list.
2. Clarify: Make sure you have a common understanding of what the item means or represents.
3. Identify three possible actions that will solve or eliminate the barrier.
4. Prioritize the possible solutions: Plan A, Plan B, Plan C.
5. Focus on what *will* work; be positive and constructive.
6. Implement the solutions; report back; celebrate and extend the ones that are effective.

Caveat: During implementation of group work, expect some challenges, barriers, and problems. Doing so will help you recognize a roadblock when it appears. When it does appear, apply the appropriate parts of Step 2 above.

With one or more colleagues, develop three or more solutions. Implement one, evaluate, replan, and retry.

The problem identification/problem formulation/problem solving format described above does not guarantee that your teamwork experiences will be free of troubles. But having a format for getting problems out on the table and then dealing with them in a problem-solving manner usually reduces the frustration and interference of group problems.

Reflection: Teamwork

I've tried to address many of the highlights of effective teamwork and team problem solving, but I've barely scratched the surface. Hundreds of books and articles have been written on effective teamwork, and I've listed a few of my favorites in the reference section (in particular, see Fisher, Rayner, and Belgard, 1995; Goldberg, 1995; Hackman, 1990; Katzenbach and Smith, 1993a, 1993b). As

I mentioned earlier, the most widely used teamwork book is Scholtes, Joiner, and Streibel's *The Team Handbook* (1996).

Questions

1. What other skills do you feel are essential for successful groups? How about trust and loyalty, for example? I briefly dealt with trust and loyalty in the section "Organization," but you may want to emphasize them more. Check the references for more (e.g., see Johnson and F. Johnson, 1991). What other teamwork skills would you like to follow up on?
2. What are some of the strategies for developing a good set of working conditions in a group?
3. What are your reactions to the list of characteristics of effective leaders in Table 3.3? Were you surprised by the high ranking of honesty?
4. Why is conflict central to effective teamwork and project work? What are some strategies for effectively managing conflict?
5. Keep a log of problems you've faced in working on project teams. How do the problems change over the life of the group?
6. The next time a problem occurs in a group, try the problem-solving process outlined in this chapter. Then evaluate how well it worked.

Exercises

1. They'll Never Take Us Alive!!

Below, in alphabetical order, are listed the top 15 causes of death in the United States in 1997. The data were taken from an annual review of death certificates. Your task is to rank them in the order of the number of deaths each causes each year. Place the number 1 next to the product or activity that causes the most deaths, the number 2 by the one that causes the second most deaths, and so on. Then, in the last column, write in your estimate of the number of fatalities each product or activity causes annually.

Product or Activity	Ranking	Number of Fatalities
Accidents		
Alzheimer's disease		
Blood poisoning		
Cancer		
Diabetes		
Hardening of arteries		
Heart disease		
HIV and AIDS		
Homicide		
Kidney disease		
Liver disease		
Lung disease		
Pneumonia and influenza		
Stroke		
Suicide		

Group Tasks

1. After individuals have filled in their charts, determine one ranking for the group. (Don't worry yet about the estimates for the numbers of fatalities.)
2. Every group member must be able to explain the rationale for the group's ranking.
3. When your group finishes, and each member has signed the chart, (a) record your estimated number of fatalities in the United States for each, and then (b) compare your ranking and estimates with those of another group.

Note: A list of rankings and annual fatalities is available from the author at ksmith@umn.edu.

2. How I Act in Conflict

The proverbs listed below can be thought of as descriptions of some of the different strategies for resolving conflicts. Proverbs state conventional wisdom, and the ones listed here reflect traditional wisdom for resolving conflicts. Read each carefully. Using the scale provided, indicate how typical each proverb is of your actions in a conflict. Then score your responses on the chart at the end of the exercise. The higher the total score in each conflict strategy, the more frequently you tend to use that strategy. The lower the total score for each conflict strategy, the less frequently you tend to use that strategy.

5 = Very typical of the way I act in a conflict
4 = Frequently typical of the way I act in a conflict
3 = Sometimes typical of the way I act in a conflict
2 = Seldom typical of the way I act in a conflict
1 = Never typical of the way I act in a conflict

______ **1.** It is easier to refrain from quarreling than to retreat from a quarrel.
______ **2.** If you cannot make a person think as you do, make him or her do as you think.
______ **3.** Soft words win hard hearts.
______ **4.** You scratch my back, I'll scratch yours.
______ **5.** Come now and let us reason together.
______ **6.** When two quarrel, the person who keeps silent first is the most praiseworthy.
______ **7.** Might overcomes right.
______ **8.** Smooth words make smooth ways.
______ **9.** Better half a loaf than no bread at all.
______ **10.** Truth lies in knowledge, not in majority opinion.
______ **11.** He who fights and runs away lives to fight another day.
______ **12.** He hath conquered well that hath made his enemies flee.
______ **13.** Kill your enemies with kindness.
______ **14.** A fair exchange brings no quarrel.
______ **15.** No person has the final answer but every person has a piece to contribute.
______ **16.** Stay away from people who disagree with you.
______ **17.** Fields are won by those who believe in winning.
______ **18.** Kind words are worth much and cost little.
______ **19.** Tit for tat is fair play.
______ **20.** Only the person who is willing to give his or her monopoly on truth can ever profit from the truths that others hold.
______ **21.** Avoid quarrelsome people as they will only make your life miserable.

______**22.** A person who will not flee will make others flee.
______**23.** Soft words ensure harmony.
______**24.** One gift for another makes good friends.
______**25.** Bring your conflicts into the open and face them directly; only then will the best solution be discovered.
______**26.** The best way of handing conflicts is to avoid them.
______**27.** Put your foot down where you mean to stand.
______**28.** Gentleness will triumph over anger.
______**29.** Getting part of what you want is better than not getting anything at all.
______**30.** Frankness, honesty, and trust will move mountains.
______**31.** There is nothing so important that you have to fight for it.
______**32.** There are two kinds of people in the world, the winners and the losers.
______**33.** When someone hits you with a stone, hit back with a piece of cotton.
______**34.** When both people give in halfway, a fair settlement is achieved.
______**35.** By digging and digging, the truth is discovered.

Scoring

Withdrawal	**Forcing**	**Smoothing**	**Compromise**	**Confrontation**
1.	2.	3.	4.	5.
6.	7.	8.	9.	10.
11.	12.	13.	14.	15.
16.	17.	18.	19.	20.
21.	22.	23.	24.	25.
26.	27.	28.	29.	30.
31.	32.	33.	34.	35.
Total	Total	Total	Total	Total

Source: David Johnson and Frank Johnson, 1991.

3. Case Study—Ralph Springer

The following case gives you a chance to apply the Blake and Mouton (1964) conflict model to a hypothetical situation. Read the case carefully and then label each of the possible actions from most to least effective and from most to least likely.

You have been working as a project manager in a large company for some time. You are friends with most of the other project managers and, you think, respected by all of them. A couple of months ago, Ralph Springer was hired as a supervisor. He is getting to know the other project managers and you. One of the project managers in the company, who is a friend of yours, confided in you that Ralph has been saying rather nasty things about your looks, the way you dress, and your personal character. For some reason you do not understand, Ralph has taken a dislike to you. He seems to be trying to get other project managers to dislike you also. From what you hear, there is nothing too nasty for him to say about you. Your are worried that some people might be influenced by him and that some of your co-project managers are also beginning to talk about you behind your back. You are terribly upset and angry at Ralph. You have a good job record and are quite skilled in project management, so it would be rather easy for you to get another job.

Rank each of the following five courses of action from 1 (most effective, most likely) to 5 (least effective, least likely). Use each number only once. Be realistic.

Effective Likely

_____ _____ I lay it on the line. I tell Ralph I am fed up with the gossip. I tell him that he'd better stop talking about me behind my back, because I won't stand for it. Whether he likes it or not, he is going to keep his mouth shut about me or else he'll regret it.

_____ _____ I try to bargain with him. I tell him that if he will stop gossiping about me I will help him get started and include him in the things other project managers and I do together. I tell him that others are angry about the gossiping and that it is in his best interest to stop. I try to persuade him to stop gossiping in return for something I can do.

_____ _____ I try to avoid Ralph. I am silent whenever we are together. Whenever we speak, I show a lack of interest, look over his shoulder, and get away as soon as possible. I want nothing to do with him for now. I try to cool down and ignore the whole thing. I intend to avoid him completely if possible.

_____ _____ I call attention to the conflict between us. I describe how I see his actions and how it makes me feel. I try to begin a discussion in which we can look for a way for him to stop making me the target of his conversation and a way to deal with my anger. I try to see things from his viewpoint and seek a solution that will suit us both. I ask him how he feels about my giving him this feedback and what his point of view is.

_____ _____ I bite my tongue and keep my feelings to myself. I hope he will find out that the behavior is wrong without my saying anything. I try to be extra nice and show him that he's off base. I hide my anger. If I tried to tell him how I feel, it would only make things worse.

4. Group Ground Rules Contract Form

Project groups are an effective aid to learning, but to work best they require that all groups members clearly understand their responsibilities to one another. These project group ground rules describe the general responsibilities of every member to the group. You can adopt additional ground rules if your group believes they are needed. Your signature on this contract form signifies your commitment to adhere to these rules and expectations.

All group members agree to:

1. Come to class and team meetings on time.
2. Come to class and team meetings with assignments and other necessary preparations done.

Additional ground rules:

1.

2.

If a member of the project team repeatedly fails to meet these ground rules, other members of the group are expected to take the following actions:

Step 1: (fill in this step with your group)
    If not resolved:
Step 2: Bring the issue to the attention of the teaching team.
    If not resolved:
Step 3: Meet as a group with the teaching team.

The teaching team reserves the right to make the final decisions to resolve difficulties that arise within the groups. Before this becomes necessary, the team should try to find a fair and equitable solution to the problem.

Member's Signatures: Group Number:____________

1. ____________________ 2. ____________________

3. ____________________ 4. ____________________

Source: Adapted from a form developed by Dr. Deborah Allen, University of Delaware.

References

Blake, R. R., and J. S. Mouton. 1964. *The managerial grid.* Houston: Gulf.

Collins, Jim. 2001a. *Good to great: Why some companies make the leap . . . and others don't.* New York: Harper Business.

———. 2001b. Level 5 leadership: The triumph of humility and fierce resolve. *Harvard Business Review,* 67–76.

Fisher, Kimball, Steven Rayner, and William Belgard. 1995. *Tips for teams: A ready reference for solving common team problems.* New York: McGraw-Hill.

Garvin, David, and Michael Roberto. 2001. What you don't know about making decisions. *Harvard Business Review* 79(8): 108–116.

Goldberg, David E. 1995. *Life skills and leadership for engineers.* New York: McGraw-Hill.

Hackman, J. R. 1990. *Groups that work (and those that don't): Creating conditions for effective teamwork.* San Francisco: Jossey-Bass.

Johnson, David W., and Frank Johnson. 1991. *Joining together: Group theory and group skills.* Upper Saddle River, NJ: Prentice Hall.

Johnson, David W., Roger T. Johnson, and Karl A. Smith. 1998. *Active learning: Cooperation in the college classroom.* (Second Edition). Edina, MN: Interaction Book Company.

Johnson, David W., Roger T. Johnson, and Karl A. Smith. 2000. Constructive controversy: The power of intellectual conflict. *Change* 32(1), 28–37.

Katzenbach, Jon R., and Douglas K. Smith. 1993a. The discipline of teams. *Harvard Business Review* 71(2): 111–120.

———. 1993b. *The wisdom of teams: Creating the high-performance organization.* Cambridge, MA: Harvard Business School Press.

Kaufmann, Walter. 1973. *Without guilt and justice: From decidophobia to autonomy.* New York: Peter H. Wyden.

Kouzes, J. M., and B. Z. Posner. 1987. *The leadership challenge: How to get extraordinary things done in organizations.* San Francisco: Jossey-Bass.

———. 1993. *Credibility: How leaders gain and lose it, why people demand it.* San Francisco: Jossey-Bass.

McNeill, Barry, Lynn Bellamy, and Sallie Foster. 1995. *Introduction to engineering design.* Tempe: Arizona State University.

Napier, Rodney W., and Matti K. Gershenfeld. 1973. *Groups: Theory and experience.* Boston: Houghton Mifflin.

Pausch, Randy. 2002. *Tips for working successfully in a group.* http://wonderland.hcii.cs.cmu.edu/Randy/teams.htm (accessed 1/4/03).

Russo, J. Edward, and Paul J. H. Shoemaker. 2002. *Winning decisions.* New York: Currency Doubleday.

Scholtes, Peter R. 1998. *The leader's handbook: Making things happen, getting things done.* New York: McGraw-Hill.

Scholtes, Peter R., Brian L. Joiner, and Barbara J. Streibel. 1996. *The team handbook,* 2nd ed. Madison, WI: Joiner Associates.

Schrage, Michael. 1991. *Shared minds.* New York: Random House.

———. 1995. *No more teams! Mastering the dynamics of creative collaboration.* New York: Doubleday.

Taylor, James. 1998. *A survival guide for project managers.* New York: AMACOM.

CHAPTER 4

Project Management Principles and Practices

This chapter discusses what a project is, introduces project-scoping strategies, and explains why projects and project management are receiving a lot of attention right now. The number of books and articles on project management is growing almost exponentially. Something is happening here. Perhaps it is due in part to observations which indicate that only a fraction of technology projects in the United States finish on time. In the United States, we spend more than $250 billion each year on IT application development of approximately 175,000 projects. . . . A great many of these projects will fail. The Standish Group research shows a staggering 31.1% of projects will be canceled before they ever get completed. Further results indicate 52.7% of projects will cost 189% of their original estimates (Standish Group, 1995).

The situation has changed a lot since the development of scheduling tools and strategies such as CPM (critical path method) and PERT (program evaluation and review technique) in the 1950s. Laufer, Denker, and Shenhar (1996) have outlined the evolution in the nature of project management. A summary of the changes is shown in Table 4.1. Laufer et al. emphasize that projects have

Table 4.1 Evolution of Models of Project Management

Central Concept	Era of Model	Dominant Project Characteristics	Main Thrust	Metaphor	Means
Scheduling (control)	1960s	Simple, certain	Coordinating activities	Scheduling regional flights in an airline	Information technology, planning specialists
Teamwork (integration)	1970s	Complex, certain	Cooperation between participants	Conducting a symphony orchestra	Process facilitation, definition or roles
Reducing uncertainty (flexibility)	1980s	Complex, uncertain	Making stable decisions	Exploring an unknown country	Search for information, selective redundancy
Simultaneity (dynamism)	1990s	Complex, uncertain, quick	Orchestrating contending demands	Directing a three-ring circus with continuous program modification based on live audience feedback	Experience, responsiveness, and adaptability

become more complex and the time to accomplish them has become shorter; thus, many projects require simultaneous management.

> REFLECTION Take a moment to reflect on the current nature of project management from your experience. Think about the project and project management environment you will enter when you graduate (especially if you're close to graduation). Try to complete the categories in Table 4.1—central concept, dominant project characteristics, main thrust, metaphor, and means.

I ask students in many of my project management classes to complete this reflection exercise. I know it is difficult to predict the future (to paraphrase Yogi Berra and Niels Bohr), but thinking about the kind of project management future we want is important, too. As Alan Kay (1971) said, "The best way to predict the future is to invent it." Alan Kay worked at Xerox Palo Alto Research Lab from 1972 to 1983 where he coined the term "Object Oriented Programming," invented the Smalltalk programming language, conceived the laptop computer and was the architect of the modern windowing GUI (graphical user interface). Here's what a couple recent groups came up with:

Central concept—Virtual, nonlinear
Dominant project characteristics—Leverage chaos
Main thrust—Melding innovation
Metaphor—Adhocracy
Means—Open-ended management

Central concept—Global projects
Dominant project characteristics—Cross-time, cross-geography
Main thrust—Parallelism
Metaphor—Concurrent engineering on steroids
Means—Electronic sharing

Ed Yourdon claims in his book *Death March Projects* (1997) that many projects must be completed in half the time, with half the budget, or with half the resources initially planned, hence the phrase "death march projects." Yourdon also claims that it is almost exciting at times to be a part of this type of project.

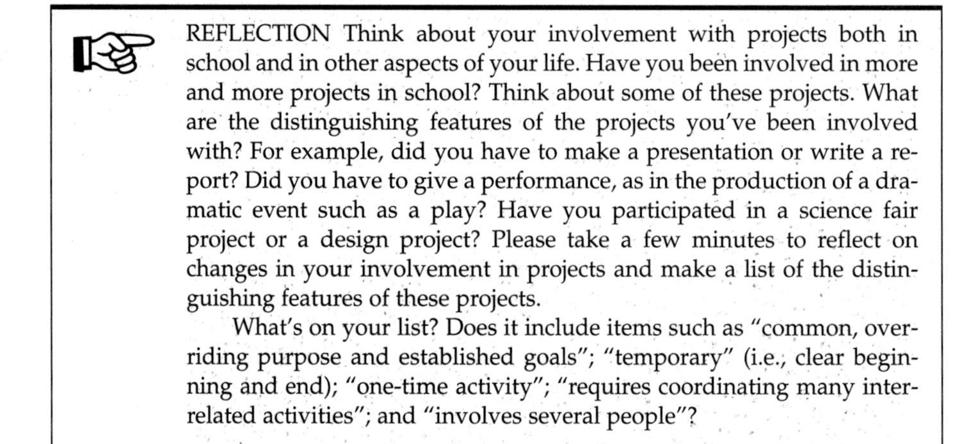

REFLECTION Think about your involvement with projects both in school and in other aspects of your life. Have you been involved in more and more projects in school? Think about some of these projects. What are the distinguishing features of the projects you've been involved with? For example, did you have to make a presentation or write a report? Did you have to give a performance, as in the production of a dramatic event such as a play? Have you participated in a science fair project or a design project? Please take a few minutes to reflect on changes in your involvement in projects and make a list of the distinguishing features of these projects.

What's on your list? Does it include items such as "common, overriding purpose and established goals"; "temporary" (i.e., clear beginning and end); "one-time activity"; "requires coordinating many interrelated activities"; and "involves several people"?

What Is a Project?

A dictionary of project management terms (Cleland and Kerzner, 1985) defines a project as follows:

> [A project is] a combination of human and nonhuman resources pulled together in a temporary organization to achieve a specified purpose.

Project is defined by Snead and Wycoff (1997) as "a nonroutine series of tasks directed toward a goal" (p. 10). In their helpful guide, the authors claim that "success depends on the ability to effectively complete projects" (p. 11).

A textbook (Nicholas, 1990) that I have used in my project management classes lists the following features of projects:

- Definable purpose with established goals
- Cost, time, and performance requirements
- Multiple resources across organizational lines
- One-time activity
- Element of risk
- Temporary activity
- Process of phases/project life cycle

Based on this definition and list of features, you can see that projects are quite different from the ongoing, day-to-day work that most of us do. Each project is unique, is temporary, has an element of risk, and has a definable purpose with established goals. Three features of projects that I'd like to explore further are (1) exploration versus exploitation projects, (2) cost, time, and performance requirements, and (3) project phases or life cycle.

Exploration Versus Exploitation Projects

A common tension I've encountered working with students on project management topics is between (1) doing old things better and (2) doing new things. Civil engineering construction projects often involve improving the effectiveness and efficiency of current practices, whereas mechanical engineering product design projects emphasize developing new ideas. James March's (1991) distinction between exploration and exploitation activities, summarized in Table 4.2, provides some guidance on the differences.

Table 4.2 Exploiting Old Ways Versus Exploring New Ways

Exploiting Old Ways: Organizing for Routine Work	**Exploring New Ways: Organizing for Innovative Work**
Drive our variance	Enhance variance
See old things in old ways	See old things in new ways
Replicate the past	Break from the past
Goal: Make money now	Goal: Make money later

Source: March, 1991.

Keys to Project Success

Traditionally, project success has been measured according to three criteria: cost, time, and performance. Although students in classes often negotiate time (especially due dates) and performance requirements, there is often less flexibility in professional life. For example, the due dates for submitting research proposals to funding agencies are rigid. One must get the proposal in before the deadline or wait until next year (and hope the agency still is making grants in that particular area). In many large construction projects there are significant incentives for finishing on time, and major penalties for finishing late. Some projects have been terminated when there were cost overruns—note the tragic demise of the Superconducting Supercollider (the multibillion-dollar particle accelerator in Texas that was terminated by the U.S. Congress).

Subsequent chapters of this book will explore how cost, time, and performance are operationalized—that is, how they are put into practice. Briefly, cost is operationalized by budgets, time by schedules, and performance by specifications and requirements.

Cost, time, and performance. Is this it? Is this all that we need to attend to for successful projects? Many project management experts are discussing a fourth aspect of project success—client acceptance. Pinto and Kharbanda (1995), for example, argue that there is this quadruple constraint on project success, which of course increases the challenge of completing projects successfully (see Figure 4.1).

The most common way of operationalizing client acceptance is by involving the client throughout the project. One of the most famous examples of this is Boeing's 777 project, in which customers were involved early on and throughout the project. These customer airlines had a significant influence on

Figure 4.1 Project Success: Quadruple Constraint.

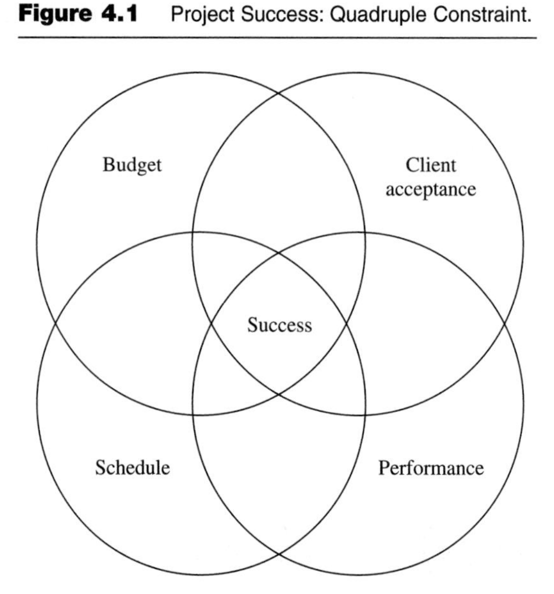

how the 777 was designed and built. Boeing's vision was to build a high-quality aircraft in an environment of no secrecy and no rivalry. These new values were clarified in the following three statements (cited in Snead and Wycoff, 1997, p. 59):

1. Use a style of management unheard of in the industry: working together while building trust, honesty, and integrity.
2. In the past, people were afraid to state a problem because of the practice of killing the messenger. We will instead celebrate our problems and get them out into the open so we can work on them.
3. We must come with no limitations in our mind. We must have a shared thought, vision, appreciation, and understanding of what we are going to accomplish together.

The creation of the environment described above was enabled by Boeing's long-range goals for the 777: (pp. 59–60)

- Design, develop, and produce a plane safer and more reliable than any other plane in aviation history that is state-of-the-art and service-ready on delivery, to be called the 777.
- Design, develop, and produce a program to empower a massive team of people to implement the "working together" philosophy while creating the 777.

Phil Condit, Boeing's CEO, said, "The task for us at Boeing is to provide a massive change in thinking throughout the company—this is a cultural shift, and it isn't easy!" Boeing experienced many positive changes (and outcomes) during this process. The 777 was delivered on time and under budget. Most significantly, however, the process positively changed the "management-teamwork" paradigm from a hierarchical relationship to a lateral relationship." (cited in Snead and Wycoff, 1997, p. 59)

If you'd like to explore Boeing's 777 project in more detail, the book *21st Century Jet* by Karl Sabbagh (1996) and the (1995) six-part PBS video series *21st Century Jet: The Building of the 777* provide rich insight into the process. Also, Jim Lewis's (2000) book *Working Together* is based on the twelve guiding principles of project management that Alan Mulally, President of Commercial Airlines Division of Boeing, used in the Boeing 777 project.

Project Life Cycle

> ☞ REFLECTION Please reconsider the projects that came to mind during the Reflection on page 48. Did each project seem to go through a series of stages? If so, how would you characterize them? Think about how the activities and work on the project changed from beginning to end. Jot down your reflections.

The prevailing view of the project life cycle is that projects go through distinct phases, such as these:

1. Conceiving and defining the project
2. Planning the project
3. Implementing the plan
4. Completing and evaluating the project
5. Operating and maintaining project

A typical construction project has the following seven phases (Kerzner, 1998):

1. Planning, data gathering, and procedures
2. Studies and basic engineering
3. Major review
4. Detail engineering
5. Detail engineering/construction overlap
6. Construction
7. Testing and commissioning

Some people, however, perhaps in moments of frustration, have described the phases of a project in more cynical terms:

1. Wild enthusiasm
2. Disillusionment
3. Total confusion
4. Search for the guilty
5. Punishment of the innocent
6. Praise and honors for the nonparticipants

These faults could often be avoided if project managers would think about resource distribution over the project life cycle.

I usually use a hands-on building exercise to help students experience project management throughout the life cycle. I give teams an in-class project of building a simply supported newprint beam that will support a concentrated load of 250 grams in the center of a span that is at least 65 centimeters long. The students find it very challenging, but most succeed—as shown in Figure 4.2. As you examine the three paper beam design images in Figure 4.2, think about important heuristics for managing a successful paper beam design project.

Figure 4.2 Simply Supported Beam Building Exercise.

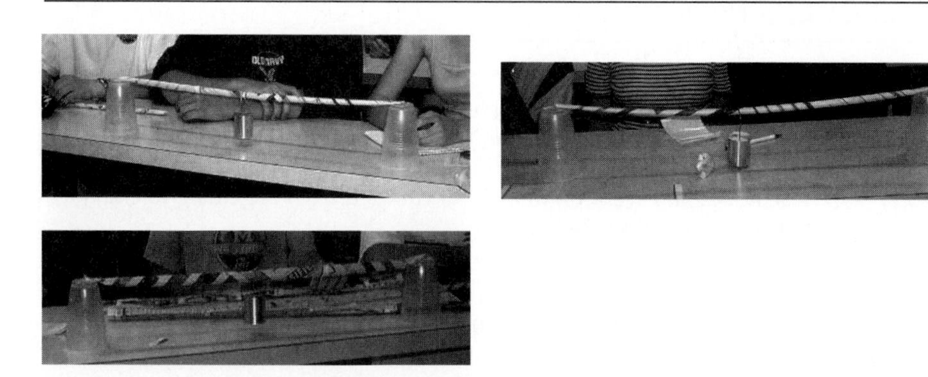

> ☞ REFLECTION Consider the paper beam project and the first four phases of the project life cycle described above (conceiving, planning, implementing, and completing) and sketch a graph of how you think resources (people, money, etc.) are distributed throughout the life of a project.
>
> What did you come up with? Continually increasing resources? Increasing, then decreasing? Why did you draw the shape of graph you did? A typical distribution of resources is shown in Figure 4.3.

Project managers must also consider how their ability to make changes and the cost of those changes vary over the project life cycle. Figure 4.4 shows the relationship between these two factors. Consider the essential message in Figure 4.4: You have considerably more flexibility early in a project and it's cheaper to make changes then, so don't skimp on planning during the early stages. This essential

Figure 4.3 Resource Distribution over the Project Life Cycle.

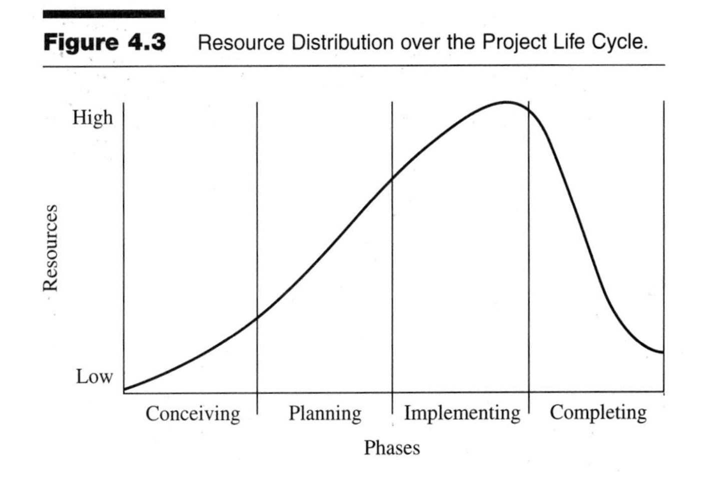

Figure 4.4 Ability to Change, and Cost to Make Changes, over the Project Life Cycle.

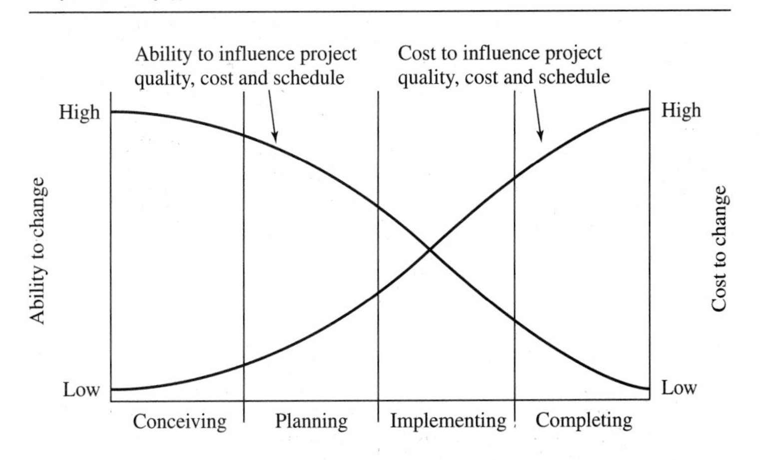

message probably makes a lot of sense, but it's hard to implement. Many project managers and project team members have such extraordinary eagerness to "get going," they often neglect to plan carefully and thoroughly. This essential message could also be described as a project management heuristic. (See Chapter 1 for elaboration on the meaning of *heuristics* and the importance of heuristics in engineering.) Following are examples of project management heuristics that students typically synthesize from the simply supported beam building exercise:

1. *Allocate resources to the weak link.* Students are more likely to succeed if they recognize that the beam usually fails at the point where the load is applied and at the ends, and apply reinforcement (their one file folder label) to these areas.
2. *Freeze the design.* At some stage in the project (when about 75 percent of the time or resources are used up) the design must be frozen. Students who individually strive to create successful prototypes without comparing designs with one another or discussing strategy often fail to create a working design as a group—they are still working individually when time is called.
3. *Periodically discuss the process and ask meta-level questions, such as, What are we doing? Why are we doing it? How does it help?* Students who reflect out loud with one another during the design process produce better designs more quickly than those who don't.

Heuristics are essential for successful project management, because every project is unique and requires its own approach. Ravindran, Phillips, and Solberg (1987) present a superb collection of modeling heuristics (that are highly relevant for project management):

1. Do not build a complicated model when a simple one will suffice.
2. Beware of molding the problem to fit the technique.
3. The deduction phase of modeling must be conducted rigorously.
4. Models should be validated prior to implementation.
5. A model should never be taken too literally.
6. A model should neither be pressed to do, nor criticized for failing to do, that for which it was never intended.
7. Beware of overselling a model.
8. Some of the primary benefits of modeling are associated with the process of developing the model.
9. A model cannot be any better than the information that goes into it.
10. Models cannot replace decision makers.

The heuristics given in both of the above lists are important when thinking about the project life cycle, and they will become crucially important when we look at the use of project scheduling models in Chapter 6.

Project Planning

Projects typically start with at statement of work (SOW) provided by the client. The statement of work is a narrative description of the work required for the project. In engineering classes, the statement of work is provided by the faculty member.

Project planning starts in response to the statement of work. This process can be very detailed, as in Mantel, Meredith, Shafer, and Sutton's (2001) project master plan, which has the following elements:

- Overview
    - Brief description of project
    - Deliverables
    - Milestones
    - Expected profitability and competitive impact
    - Intended for senior management
- Objectives
    - Detailed description of project's deliverables
    - Project mission statement
- General Approach
    - Technical and managerial approaches
    - Relationship to other projects
    - Deviations from standard practices
- Contractual Aspects
    - Agreements with clients and third parties
    - Reporting requirements
    - Technical specifications
    - Project review dates
- Schedules
    - Outline of all schedules and milestones
- Resource Requirements
    - Estimated project expenses
    - Overhead and fixed charges
- Personnel
    - Special skill requirements
    - Necessary training
    - Legal requirements
- Evaluation Methods
    - Evaluation procedures and standards
    - Procedures for monitoring, collecting, and storing data on project performance
- Potential Problems
    - List of likely potential problems

Another common format for project planning is the project charter, and in this area there has been a radical change recently. I've asked participants in executive project management workshops over the years how they go about planning a project. Five years ago no one mentioned project charters, but in the past 2 or 3 years almost every participant has had experience with sophisticated project-planning systems. They also treat their systems as intellectual property and are reluctant to share them.

Martin and Tate (1997) describe the following elements in a typical project charter:

1. Write an overview of the project scope.
2. Determine the team's boundaries for creating the deliverables.
3. Define the customer's criteria for acceptance.
4. Determine the required reviews and approvals.
5. Establish risk limits.
6. Select the project leader and team members.
7. Set deadlines for delivery of the final deliverables.
8. Set limits on staffing and spending.
9. Create a list of required reports.
10. Identify organizational constraints and project priorities.

The next level of detail in the planning process is the development of a work breakdown structure (WBS). A WBS is "a deliverable-oriented grouping of project elements which organizes and defines the total scope of a project" (Duncan, 1996). Typically a WBS has three to six levels, such as program, project, task, and subtask. Developing a work breakdown structure is important for scoping a project—that is, determining the specific tasks that have to be completed, choosing appropriate groupings for these activities, and setting precedence and interdependence (what has to follow what, and what can be going on at the same time).

A simple approach for creating a WBS is to (1) gather the project team, (2) provide team members with pad of sticky notes, (3) ask team members to write down all the tasks they can think of, (4) have team members place their sticky notes on chart paper, and (5) have members work together to rearrange them. I've used the following Post-It Note Project Planning exercise in my classes and workshops:

Statement of Work (SOW)—Activities Needed to Complete Office Remodeling

1. One activity per Post-It note. Include name, description, and estimated duration (initial each Post-It).
2. Arrange Post-Its on chart paper.
3. Work together to rearrange Post-Its.
4. Draw arrows to indicate precedence.

Figure 4.5 shows typical results of this activity.

Figure 4.5 Post-It Note Project Planning.

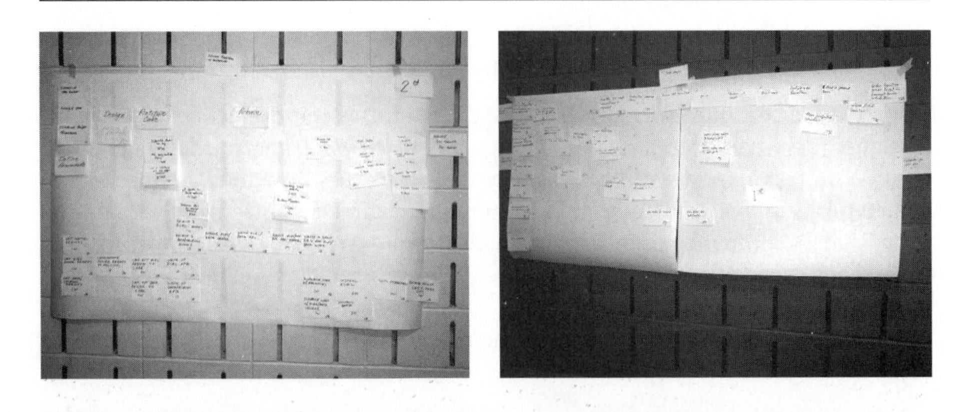

These two parts of project planning in response to the statement of work, the Project Charter and work breakdown structure—are often neglected in traditional project management textbooks and classes, perhaps due to an eagerness to get to the nitty-gritty of doing project scheduling using critical path analysis. However, carefully considering these two initial aspects of project scoping is an important part of thorough planning.

Reflection: Project Management

Project management is a relatively new profession and is growing at a remarkable rate. *Fortune* magazine called project management "Career Number 1" for the 1990s. When I was in engineering school in the late 1960s, project management courses weren't offered. Because I now teach several project management courses, I've had to learn it by experience and research. Several of the books I've found useful and have used as texts in my courses are listed in the references (see Culp and Smith, 1992; Eisner, 1997; Frame, 1994, 1995; Grady, 1992; Graham and Englund, 1997; Kerzner, 1998; Lewis, 1993, 1998, 2001; Lientz and Rea, 1995; Meredith and Mantel, 1994; Nicholas, 1990; Pinto and Kharbanda, 1995; Snead and Wycoff, 1997). Because project management is an emerging field and is changing quite rapidly, I encourage you to continue honing your skills and competencies.

Questions

1. What is a project? What are the key characteristics of projects? How does project management differ from management in general?
2. What are the three cardinal conditions of project success?
3. What has been your experience in completing projects "on time, under budget, within specifications, and acceptable to the client"? What is your "batting average"? Is it better than the 10 percent figure cited by the Project Management Institute survey?
4. How has your distribution of effort varied over the life of projects that you've worked on? Do you start strong and taper off? Do you start slowly and build? Sketch out a graph that has effort on the ordinate (y) and time on the abscissa (x) for a typical project. Is your effort curve consistent with the "bell-shaped" curve described earlier in the chapter? Is it different? Does most of your effort happen during the last few hours before the project is due? If much of your effort is applied in the closing hours of the project, perhaps you are freezing the design too late. How does your enthusiasm vary over the project life cycle?
5. Start developing a list of your own "project management heuristics." There are several books of rules of thumb, which are one type of heuristic. One of my favorites is Grady (1992).

Exercises

1. Project Planning

Now that you've had an opportunity to think about projects, project management, project life cycles, and project scoping, I'd like you to try applying what you've learned.

Suppose you have two tickets to a fabulous concert and are planning a special dinner for two prior to the concert. Your menu consists of a very special soup and a baked chicken entrée. The soup must be boiled for 35 minutes and you should allow 15 minutes to serve and consume it. The chicken dish requires a fair amount of preparation: you have to boil the rice for 30 minutes, brown the chicken in the frying pan for 15 minutes, and bake the rice and chicken in a baking dish in the oven for 15 minutes. It takes 5 minutes to prepare a sauce in the frying pan and 15 minutes to boil the peas. (You have only two pots and one frying pan.) For the good red wine you've bought, you must allow 5 minutes to uncork it (very carefully) and let it stand for 30 minutes before serving it. You plan to allow 25 minutes to serve and eat the entrée and drink the wine.

How much time do you need to prepare and consume the meal? What representation (model) did your group use to determine the time? How did you keep track of which activities had to follow others and which could be going on at the same time?

2. Playground Project Planning

Student professional organizations (American Society of Civil Engineering, ASCE) are often involved in community service projects, such as the following school playground construction project. Assume that you are a project team member and that the following activities must be accomplished to build a playground.

Activity	Estimated Duration (Quarter Days)
Construct playground pads and slides	1
Construct playground walls	2
Design playground	3
Drill foundation holes	1
Excavate site	1
Place sand	1
Place support frame	2
Procure cement	5
Procure playground materials	7
Procure tools	1
Wash playground	1

You have been assigned by the project manager to complete the following tasks:

Deliverables:

1. Develop a brief statement of work or project mission statement.
2. Complete a work breakdown structure.

3. General Project Scoping

Select a project (or subproject) from your workplace or experience, and assume the following roles and prepare the appropriate documents.

Project Deliverables

1. Assume the role of client: Develop a statement of work (SOW) or project mission statement. Write the SOW as the client directing the engineer-consultant-contractor to perform prescribed project details and requirements.
2. Assume the role of engineer/consultant/contractor. Create a project charter using the Martin and Tate (1997) format (see page 56), the Lewis Institute charterform (www.lewisinstitute.com/pdf/charterform.pdf), or a form from your experience or workplace.

3. Complete a work breakdown structure (WBS) for the project.
4. Write a one-page executive summary of the project.

References

Cleland, D. I., and H. Kerzner. 1985. *A project management dictionary of terms.* New York: Van Nostrand Reinhold.

Culp, G., and A. Smith. 1992. *Managing people (including yourself) for project success.* New York: Van Nostrand Reinhold.

Duncan, William R. 1996. *A guide to the project management body of knowledge.* Newton Square, PA: Project Management Institute.

Eisner, H. 1997. *Essentials of project management and systems engineering management.* New York: Wiley.

Frame, J. D. 1994. *The new project management.* San Francisco: Jossey-Bass.

———. 1995. *Managing projects in organizations.* San Francisco: Jossey-Bass.

Grady, Robert B. 1992. *Practical software metrics for project management and process improvement.* Englewood Cliffs, NJ: Prentice Hall.

Graham, Robert J., and Randall L. Englund. 1997. *Creating an environment for successful projects.* San Francisco: Jossey-Bass.

Kay, Alan. 1971. The origin of the quote came from an early meeting in 1971 of PARC, *Palo Alto Research Center,* folks and the Xerox planners. "In a fit of passion I uttered the quote!"—Alan Kay, in an e-mail on Sept. 17, 1998 to Peter W. Lount (http://www.smalltalk.org/alankay.html, accessed 3/9/03).

Kerzner, H. 1998. *Project management: A systems approach to planning, scheduling, and controlling,* 6th ed. New York: Van Nostrand Reinhold.

Laufer, A., G. R. Denker, and A. J. Shenhar. 1996. *Simultaneous management: The key to excellence in capital projects.* International Journal of Project Management 14(4): 189–199.

Lewis, James P. 1993. *The project manger's desk reference: A comprehensive guide to project planning, scheduling, evaluation, control and systems.* New York: Probus.

Lewis, James P. 1995. *Fundamentals of project management.* New York: AMACOM.

———. 2001. *Project planning, scheduling, and control: A hands-on guide to bringing projects in on time and on budget,* 3rd ed. New York: Probus.

Lewis, James P. 1998. *Mastering project management: Applying advanced concepts of systems thinking, control and evaluation, resource allocation.* New York: McGraw-Hill.

Lewis, James P. 2000. *Working together: 12 principles for achieving excellence in managing projects, teams, and organizations.* New York: McGraw-Hill.

Lientz, Bennet, and Kathryn Rea. 1995. *Project management for the 21st century.* San Diego: Academic Press.

March, James G. 1991. Exploration and exploitation in organizational learning. *Organizational Science* 2: 71–87.

Martin, P., and K. Tate. 1997. *Project management memory jogger.* Lawrence, MA: GOAL/QPC.

Meredith, J. R., and S. J. Mantel. 1994. *Project management: A managerial approach.* New York: Wiley.

Nicholas, J. M. 1990. *Managing business and engineering projects: Concepts and implementation.* Englewood Cliffs, NJ: Prentice Hall.

Pinto, J. K., and O. P. Kharbanda. 1995. *Successful project managers: Leading your team to success.* New York: Van Nostrand Reinhold.

Ravindran, A., D. T. Phillips, and J. J. Solberg. 1987. *Operations research: Principles and practice.* New York: Wiley.

Sabbagh, Karl. 1995. *21st century jet: The building of the 777.* PBS Home Video.

Sabbagh, Karl. 1996. *21st century jet: The making and marketing of the Boeing 777.* New York: Scribner's.

Snead, G. L., and J. Wycoff. 1997. *To do, doing, done! A creative approach to managing projects and effectively finishing what matters most.* New York: Fireside.

Standish Group. 1995. The Standish Group Report. http://www.scs.carleton.ca/~beau/PM/Standish-Report.html (accessed 3/11/03).

Yourdon, E. 1997. *Death march projects.* Reading, MA: Addison-Wesley.

CHAPTER 5

The Project Manager's Role

Project management is undergoing enormous changes, as Table 4.1 indicated, and thus the role of the project manager is changing. Before we explore the changes that are occurring in project management, let's explore broader changes that are occurring in business, industry, government, and education.

> INDIVIDUAL AND GROUP REFLECTION Think about changes that have occurred in the workplace (or school, if that is your principal area of experience) in the past 5 years. Make a list of some of the most notable changes and compare it with other team members' lists.

Students in my project management classes who do the above Reflection come up with lots of changes they're noticing—communications technology, computers, the global marketplace, emphasis on quality, shortened time frames, and the changing role and importance of knowledge workers.

Changes in the Workplace

Changes in the workplace have been studied and summarized by numerous authors, including Byrne (1992, 2000). Changes occurring in how engineers work in business and industry, summarized in Table 5.1, have serious implications for how we prepare engineering graduates for working in the 21st century.

The changes that are occurring in business and industry suggest that we should consider changes in engineering education to prepare our graduates to function effectively in the "new paradigm" companies. The "Made in America" study (Dertouzos, Lester, and Solow, 1989) recommended that MIT should do the following:

1. Broaden its educational approach in the sciences, in technology, and in the humanities, and educate students to be more sensitive to productivity, to practical problems, to teamwork, and to the cultures, institutions, and business practices of other countries.
2. Create a new cadre of students and faculty characterized by (1) interest in, and knowledge of, real problems and their societal, economic, and political context; (2) an ability to function effectively as members of a team

Table 5.1 What a Difference a Century Can Make: Contrasting Views of the Corporation

Characteristic	20th Century	21st Century
Organization	The pyramid	The web or network
Focus	Internal	External
Style	Structured	Flexible
Source of strength	Stability	Change
Structure	Self-sufficiency	Interdependencies
Resources	Atoms—physical assets	Bits—information
Operations	Vertical integration	Virtual integration
Products	Mass production	Mass customization
Reach	Domestic	Global
Financials	Quarterly	Real-time
Inventories	Months	Hours
Strategy	Top-down	Bottom-up
Leadership	Dogmatic	Inspirational
Workers	Employees	Employees and free agents
Job expectations	Security	Personal growth
Motivation	To compete	To build
Improvements	Incremental	Revolutionary
Quality	Affordable best	No compromise

Source: Byrne, 2000.

creating new products, processes, and systems; (3) an ability to operate effectively beyond the confines of a single discipline; and (4) an integration of a deep understanding of science and technology with practical knowledge, a hands-on orientation, and experimental skills and insight.

3. Revise subjects to include team projects, practical problems, and exposure to international cultures. Encourage student teaching to instill a stronger appreciation of lifelong learning and the teaching of others. Reinstitute a foreign-language requirement in the undergraduate admissions process.

Although progress has been made in the past 12 years, largely due to the Accreditation Board for Engineering and Technology (ABET) emphasis on graduates' skills and abilities, the changes Dertouzos et al. called for in 1989 are extremely important.

Changes in engineering education were described in a paper in the Frontiers in Education Conference proceedings (Smith and Waller, 1997) and are summarized in Table 5.2. If you're interested in learning more about new paradigms for engineering education, you may view the paper on the World Wide Web at the ASEE/IEEE Frontiers in Education page.

The premier issue of an exciting engineering magazine, *Today's Engineer*, proposed that we are at the dawning of a new age of engineering—the crossroads for a changing professional model (Gaynor, 1998). The editor, G. H. Gaynor, claims that this new model makes three demands on us: that we transcend traditional boundaries, that we think strategically, and that we develop a business perspective. Gaynor also tells us that technical competence is an absolute requirement, but by itself is no longer sufficient. It must be integrated with breadth of vision, flexibility, customer focus, and a business orientation.

Table 5.2 Comparison of Old and New Paradigms for College Teaching

	Old Paradigm	New Paradigm
Knowledge	Transferred from faculty to students	Jointly constructed by students and faculty
Students	Passive vessel to be filled by faculty's knowledge	Active constructor, discoverer, transformer of knowledge
Mode of learning	Memorizing	Relating
Faculty purpose	Classify and sort students	Develop students' competencies and talents
Student goals	Complete requirements, achieve certification within a discipline	Grow, focus on continual lifelong learning within a broader system
Relationships	Impersonal relationship among students and between faculty and students	Personal transaction among students and between faculty and students
Context	Competitive/individualistic	Cooperative learning in classroom and cooperative teams among faculty
Climate	Conformity/cultural uniformity	Diversity and personal esteem/cultural diversity and commonality
Power	Faculty holds and exercises power, authority, and control	Students are empowered; power is shared among students and between students and faculty
Assessment	Norm-referenced (i.e., graded "on the curve"); typically multiple-choice items; student rating of instruction at end of course	Criterion-referenced; typically performances and portfolios; continual assessment of instruction
Ways of knowing	Logico-scientific	Narrative
Epistemology	Reductionist; facts and memorization	Constructivist; inquiry and invention
Technology use	Drill and practice; textbook substitute; chalk and talk substitute	Problem solving, communication, collaboration, information access, expression
Teaching assumption	Any expert can teach	Teaching is complex and requires considerable training

Source: Smith and Waller, 1997b.

Changes like those outlined by Dertouzos and his colleagues and those boldly described by Gaynor are enormously difficult to implement in a direct, linear manner. The nature of change is described in Katzenbach and Smith (1993) using the metaphor of a whitewater raft ride. The authors also list behavioral changes that are demanded by change.

> Major change, by its nature, is intentionally disruptive and largely unprogrammable. In comparing the management of major versus normal change, one top executive said, "It used to be like I-75. You'd lay it out from Toledo to Tampa. Now it's more like a whitewater raft ride. You try to get the right people in the raft and do the best you can to steer it. But you never know what's just around the bend." (p. 208)

Katzenbach and Smith suggest several behavioral changes that will help us make the necessary changes; these are listed in Table 5.3.

Table 5.3 Behavioral Changes Demanded by Performance in the 1990s and Beyond

From	To
Individual accountability	Mutual support, joint accountability, and trust-based relationships *in addition to* individual accountability
Dividing those who think and decide from those who work and do	Expecting everyone to think, work, and do
Building functional excellence through each person executing a narrow set of tasks ever more efficiently	Encouraging people to play multiple roles and work together interchangeably on continuous improvement
Relying on managerial control	Getting people to buy into meaningful purpose, to help shape direction, and to learn
A fair day's pay for a fair day's work	Aspiring to personal growth that expands as well as exploits each person's capabilities

Peter Drucker (1993), who has written more articles for the *Harvard Business Review* than anyone else, has described the changing views of the "manager" concept. Drucker stresses the idea of the "knowledge worker" and, consistent with this concept, focuses on skills and strategies for "managing the knowledge worker." In the 1920s, a manager was seen as a person who was responsible for the work of subordinates; in the 1950s, a manager was responsible for the performance of people; and in the 1990s and beyond, a manager is responsible for the application and performance of knowledge.

INDIVIDUAL REFLECTION How are you feeling about all these impending changes? A bit overwhelmed, no doubt. Are you seeing changes in your educational experience? Is your college education on the cutting edge of modern practice? What do you think about the state of your education? Discuss with your group.

Lots of people have noted, often with a sense of frustration, that today's classroom is one of the few settings that would be recognizable to someone working in the "same" setting 100 years ago. Few, if any, other settings—transportation, hospitals, factories—would be recognizable to someone who worked in them 100 years ago. Perhaps the classroom is a "place that time forgot and that the decades can't improve" (to paraphrase Garrison Keillor from *Prairie Home Companion*), but I think there is room for improvement.

Changes in Project Management

With all the changes that are occurring in the workplace—including downsizing, rightsizing, and attending to the customer—is there any question that change is also occurring in project management? Management guru Tom Peters (1999) makes bold claims about the importance of project management: "Those organizations that take project management seriously as a discipline, as a way of life, are likely to make it into the 21st century. Those that do not are likely to find themselves in good company with dinosaurs" (p. 128). As an engineering graduate student, Peters wrote a masters thesis on PERT charts. He has also made the following statements (Peters, 1991):

- "Tomorrow's corporation is a 'collection of projects.'"
- "Everyone needs to learn to work in teams with multiple independent experts—each will be dependent upon all the others voluntarily giving their best."
- "The new lead actor/boss—the Project Manager—must learn to command and coach; that is, to deal with paradox." (p. 64)

Tom Peters's recent work is on the WOW project (Peters, 1999a). He writes in *Fast Company* (Peters, 1999b): "In the new economy, all work is project work. And you are your projects! Here's how to make them go WOW!" (p. 116).

In the area of project management, several authors have summarized the most notable changes. Pinto and Kharbanda (1995) refer to our age as "The Age of Project Management." The following are key features of this age:

1. Shortened market windows and product life cycles
2. Rapid development of third world and closed economies
3. Increasingly complex and technical products
4. Heightened international competition
5. The environment of organizational resource scarcity

Lientz and Rea (1995) list several trends that affect projects:

Global competition	Empowerment
Rapid technological change	Focus on quality and continuous
Product obsolescence	Improvement
Organizational downsizing	Measurement
Business reengineering	Interorganizational systems

Furthermore, Lientz and Rea remind us that projects are set in time. They are also set in the contexts of organization, a legal system, a political system, a technology structure, an economic system, and a social system. How do these environmental factors affect projects? How should a project manager and project respond? Outside factors impact the project, and the project must respond to the resulting challenges.

If I haven't I convinced you that there are many changes occurring in the business world and that the emergence of project management is one of them, try the following reflection.

INDIVIDUAL REFLECTION What does it take to be a good project manager? Take a few minutes to think about the skills and competencies (and perhaps the attitudes) needed for effective project management. Make a list. Compare your list with the lists of other students.

Do you know any project managers? Do you have relatives or friends who do project work? Try to find someone you can interview to help you get your bearings on project management. (See the exercise at the end of this chapter.) Then revise your list.

Skills Necessary for Effective Project Managers

Barry Posner (1987) conducted a survey of project managers, asking them what it takes to be a good project manager. He got the following results:

1. Communications (84 percent of respondents listed it)
   a. Listening
   b. Persuading
2. Organizational skills (75 percent)
   a. Planning
   b. Goal setting
   c. Analyzing
3. Team-building skills (72 percent)
   a. Empathy
   b. Motivation
   c. Esprit de corps
4. Leadership skills (68 percent)
   a. Sets example
   b. Energetic
   c. Vision (big picture)
   d. Delegates
   e. Positive
5. Coping Skills (59 percent)
   a. Flexibility
   b. Creativity
   c. Patience
   d. Persistence
6. Technological Skills (46 percent)
   a. Experience
   b. Project knowledge

Several authors have surveyed project managers and conducted extensive literature searches to learn about essential project management skills. Pinto and Kharbanda (1995) list the following skills necessary for effective project managers:

Planning: work breakdown, project scheduling, knowledge of project management software, budgeting and costing.

Organizing: team building, establishing team structure and reporting assignments, defining team policies, rules, and protocols.

Leading: motivation, conflict management, interpersonal skills, appreciation of team members' strengths and weaknesses, reward systems.

Controlling: project review techniques, meeting skills, project close-out techniques.

Lientz and Rea (1995) provide the following list of keys to success as a project manager:

Communicate regularly *in person* with key team members.
Keep management informed.
Keep informed on all aspects of the project.
Delegate tasks to team members.
Listen to input from team members.
Be able to take criticism.
Respond to and/or act on suggestions for improvement.
Develop contingency plans.
Address problems.
Make decisions.
Learn from past experience.
Run an effective meeting.
Set up and manage the project file.
Use project management tools to generate reports.

Understand trade-offs involving schedule and budget.
Have a sense of humor.

> INDIVIDUAL REFLECTION How do these lists compare with yours? Was there a lot of overlap? Were there categories of items that were on your list but not on these, and vice versa?

Research by Jeffrey Pinto (1986) sought to quantify some of these factors by correlating them with their importance for system implementation (see the box "Critical Success Factors. . . ."). "System implementation" may be interpreted as a successful project outcome.

How does one implement all of the characteristics of effective project managers? There are so many. One way is to employ a common modeling strategy, called *salami tactics,* in which a complex problem is broken into smaller, more manageable parts (Starfield, Smith, and Bleloch, 1994). The "slices" that I'll use are the phases in a typical project life cycle—planning, organizing, staffing, directing, and controlling.

Project Manager's Role over the Project Life Cycle

Planning

During the planning stage, you as the project manager must establish project objectives and performance requirements. Remember to involve key participants in the process (because, according to an old rule of thumb, involvement builds commitment). Establish well-defined milestones with deadlines. Try to

Critical Success Factors and Their Importance for System Implementation

1. *Project mission.* Initial clearly defined goals and general directions.
2. *Top management support.* Willingness of top management to provide the necessary resources and authority/power for implementation success.
3. *Schedule plans.* A detailed specification of the individual action steps for system implementation.
4. *Client consultation.* Communication, consultation, and active listening to all parties impacted by the proposed project.
5. *Personnel.* Recruitment, selection, and training of the necessary personnel for the implantation project team.
6. *Technical tasks.* Availability of the required technology and expertise to accomplish the specific technical action steps to bring the project on-line.
7. *Client acceptance.* The act of "selling" the final product to its ultimate intended users.
8. *Monitoring and feedback.* Timely provision of comprehensive control information at each stage in the implementation process.
9. *Communication.* The provision of an appropriate network and necessary data to all key actors in the project implementation process.
10. *Troubleshooting.* The ability to handle unexpected crises and deviations from plan.

Source: Pinto, 1986.

anticipate problems and build in contingencies to allow for them. Carefully outline responsibilities, schedules, and budgets.

Organizing

The first step in organizing is to develop a work breakdown structure that divides the project into units of work. If the project is large and complex, then the next step is to create a project organization chart that shows the structure and relationships of key project members. Finally, schedules, budgets, and responsibilities must be clearly and thoroughly defined.

Staffing

Most project successes depend on the people involved with the project. You must define work requirements and, to the extent possible, seek appropriate input when selecting team members. Be sure to orient team members to the big picture of the project. Seek each team member's input to define and agree upon scope, budget, and schedule. (Remember, involvement builds commitment, and usually a better product.) Set specific performance expectations with each team member.

Directing

The day-to-day directing of projects involves coordinating project components, investigating potential problems as soon as they arise, and researching and allocating necessary resources. Be sure to remember to display a positive can-do attitude, and to be available to team members. Recognize team members' good work and guide necessary improvement.

Controlling

Keeping the project on course with respect to schedule, budget, and performance specifications requires paying attention to detail. The following factors usually help:

1. Communicate regularly with team members.
2. Measure project performance by maintaining a record of planned and completed work.
3. Chart planned and completed milestones.
4. Chart monthly project costs.
5. Document agreements, meetings, telephone conversations.

This summary of the project manager's role over the project life cycle is intended to provide guidance and is not intended to be followed linearly or exhaustively. Each project is unique and requires its own approach. However, the suggestions above are likely to improve project success.

Reflection: On Change

Enormous changes are occurring in the way work and learning are done. You are probably experiencing some of these changes in your classes as you are asked to work on projects in groups and formulate and solve open-ended problems. If you're working at an engineering job, you are surely experiencing some of these changes.

I've tried to provide a perspective on changes that are occurring both in the classroom and in the workplace. One of the most influential references on change is Stephen Covey's *Seven Habits of Highly Effective People* (Covey, 1989) which has sold millions of copies. Covey lists these habits as follows:

1. *Be pro-active.* Take the initiative and the responsibility to make things happen.
2. *Begin with an end in mind.* Start with a clear destination to understand where you are now, where you're going, and what you value most.
3. *Put first things first.* Manage yourself. Organize and execute around priorities.
4. *Think win/win.* See life as a cooperative, not a competitive, arena where success is not achieved at the expense or exclusion of the success of others.
5. *Seek first to understand.* Understand then be understood to build the skills of empathic listening that inspire openness and trust.
6. *Synergize.* Apply the principles of cooperative creativity and value differences.
7. *Renewal.* Preserve and enhance your greatest asset, yourself, by renewing the physical, spiritual, mental, and social/emotional dimensions of your nature.

Countless students in my classes have said "Covey's book changed my life!" Stephen Covey had an early influence on my professional work. I attended a management development program he conducted in Moab, Utah, in 1969–1970 as a part of my first full-time engineering job. The roles of project managers in engineering school and in the workplace are complex and varied. Covey's list provides a good set of heuristics to guide project managers. Another classic that you may want to read is Frederick P. Brooks's *The Mythical Man-Month.* If Brooks's book about software project management intrigues you, then you may want to check out Steve McConnell's *Software Project Survival Guide.* More and more engineers are involved in software development, and these two books will help you manage it.

Reflection: Professor as Project Manager

I tend to think of each course I teach as a project. And in terms of beginnings and endings, and resource loading during the semester, each course clearly has project characteristics. I think about the goals and outcomes, that is, what I want students to know and be able to do at the end of the course; what evidence I'll gather through monitoring and assessment to document that students have learned; and finally I design the instructional aspects of the course. Mainly I try to orchestrate variety during the class sessions so I don't get bored and to provide help for students who learn in a myriad of ways. In the fall 2002 semester, two students wrote a paper on the project management aspects of the course. I was hesitant to read it, but as I delved in, I found from their perspective that I had succeeded in treating the course as a project. This paper by Luke Heaton and Greg Williams, *An Analysis of Project Management CE 4101W as a Project,* is available from me at ksmith@umn.edu.

Questions

1. What changes have you noted in the workplace or school? Has your school undergone a schedule change recently (e.g., from quarters to semesters)? How do the changes you've noted compare with those listed in this chapter?
2. This chapter emphasized the changing nature of the workplace and of engineering work, and the needed project manager skills, based on past (and sometimes current) practice. What do you anticipate project manager skills will be in the future? What do the futurists (John Naisbitt, Watts Walker, Esther Dyson, etc.) have to say about this?
3. What skills are essential for effective project managers? How can they be enhanced and developed?
4. What does the literature identify as keys to project success? How does this list compare with your experience?
5. This chapter organized the project manager's role around the life cycle. Are there other ways that come to mind for organizing the project manager's role? What are they? What are their advantages and disadvantages?

Exercise

Interview a project manager, or someone who is involved with project work. Potential interview questions developed by students in my project management classes are listed below. This exercise not only will give you a chance to find out more about project management in practice (and refine your list of essential skills, competencies, and attitudes), but also may help you decide if it is a career path you'd like to pursue.

1. What is the main thrill or interesting reason for being a project manager?
2. Describe a typical work schedule during the week, including number of hours.
3. What directed you to become a project manager? Why not stay in engineering or some other form of business management?
4. Discuss examples of the project manager's role as a team leader and a team member.
5. Discuss the realm and responsibilities of the project manager.
6. What personal characteristics are the project manager's best allies during the job's activities?
7. What skills have you had to develop or refine since becoming a project manager?
8. What personal goals are you striving for personally and professionally?
9. What are the things that slow you down?
10. What are a few of the frustrations of the job?
11. Describe the project manager's professional credibility and its value.
12. How does the company you work for consider the project manager and her or his responsibilities/opportunities?
13. Describe any battles you've experienced to complete tasks in the most economical manner and still maintain quality or integrity.
14. How important is the project schedule and what purposes does it serve?
15. Discuss the extent of your interaction with project owners or clients.
16. What is your level of involvement with contract negotiations?
17. Discuss your interaction with other professionals (e.g., engineer or architect).

18. Describe methods you use to monitor adequate communication between project owner, architect, client, etc., and the jobsite crew.
19. Describe the accounting practices you use for project budget and regular reviews.
20. Is your financial compensation commensurate with the work you do? What other benefits are there?

References

Brooks, Frederick P., Jr., 1995. *The mythical man-month: Essays on software engineering—Anniversary edition.* Reading, MA: Addison-Wesley.

Byrne, J. A. 1992. Paradigms for postmodern managers. *Business Week* (Special Issue on Reinventing America), pp. 62–63.

———. 2000. Management by web: The 21st century corporation. *Business Week,* August 28, p. 87.

Covey, Stephen R. 1989. *The seven habits of highly effective people.* New York: Simon & Schuster.

Dertouzos, M. L., R. K. Lester, and R. M. Solow. 1989. *Made in America: Regaining the productive edge.* New York: Harper.

Drucker, Peter F. 1993. *Post-capitalist society.* New York: Harper Business.

Gaynor, G. H. 1998. The dawning of a new age: Crossroads of the engineering profession. *Today's Engineer* 1(1): 19–22.

Katzenbach, Jon R., and Douglas K. Smith. 1993. *The wisdom of teams: Creating the high-performance organization.* Cambridge, MA: Harvard Business School Press.

Lientz, Bennet, and Kathryn Rea. 1995. *Project management for the 21st century.* San Diego: Academic Press.

McConnell, Steve. 1998. *Software project survival guide: How to ensure your first important project isn't your last.* Redmond, WA: Microsoft Press.

Oberlender, G. D. 1993. *Project management for engineering and construction.* New York: McGraw-Hill.

Peters, Thomas J. 1987. *Thriving on chaos: Handbook for a management revolution.* New York: Knopf.

Peters, Thomas J. 1991. Managing projects takes a special kind of leadership. *Seattle Post-Intelligence,* April 29, 1991, p. 64.

Peters, Thomas J. 1999a. *The Project 50.* New York: Knopf.

Peters, Thomas J. 1999b. The WOW project: In the new economy, all work is project work. *Fast Company,* 24: 116–128.

Pinto, J. K. 1986. *Project implementation: A determination of its critical success factors, moderators, and their relative importance across stages in the project life cycle.* Unpublished Ph.D. dissertation, University of Pittsburgh.

Pinto, J. K., and O. P. Kharbanda. 1995. *Successful project managers: Leading your team to success.* New York: Van Nostrand Reinhold.

Posner, Barry. 1987. Characteristics of effective project managers. *Project Management Journal* 18(1): 51–54.

Smith, Karl A., and Alisha A. Waller. 1997. *New paradigms for engineering education.* Pittsburgh: ASEE/IEEE Frontiers in Education Conference Proceedings. Adapted from K. A. Smith and A. A. Waller. After word: New paradigms for college teaching, in *New paradigms for college teaching,* edited by W. E. Campbell and K. A. Smith (Edina, MN: Interaction, 1995).

Starfield, A. M., K. A. Smith, and A. L. Bleloch. 1994. *How to model it: Problem solving for the computer age.* Edina, MN: Burgess.

CHAPTER 6

Project Scheduling

Project scheduling is a central yet often overrated aspect of project management. For some the feeling is "We've got a schedule; we're done." Getting a schedule is just one important step in the process of project management. The real work begins when circumstances cause delays and pressures mount to revise the schedule.

> INDIVIDUAL REFLECTION Think about how you typically schedule complex projects, such as completing a major report for a class. Do you make a list of things to do? an outline? Do you draw a concept map? Or do you just start writing? Do you develop a timeline?

In this chapter, we'll work our way through the details of the scheduling process. We'll learn the basics of the critical path algorithm and experience firsthand the ideas of forward pass, backward pass, critical path, and float. As you develop an understanding of these concepts and procedures, you will gain insight into managing projects with complex schedules. I first encountered these concepts as an undergraduate engineering student in about 1964. The critical path method was presented in an engineering systems class as a network algorithm. The approach was purely mechanical; no reference was made to its potential usefulness in scheduling projects.

Let's revisit the meal-planning exercise from Chapter 4 (p. 58). Look back at the work you did for this exercise. If you haven't done that exercise yet, go back and think about how you would tackle this task. We'll use this exercise as the project example throughout this chapter.

> REFLECTION What representation (model) did your group use to determine the time needed to cook and eat the meal? How did you keep track of which activities had to follow others and which could be going on at the same time? How did you go about determining the total time the meal preparation and eating would take? Did you make a list? a timeline? Or did you approach the problem in some other way?

Work Breakdown Structure

A common approach for scoping a project is to prepare a work breakdown structure (WBS) (see Chapter 4). The WBS can be presented as a list or an organization chart. A one-level WBS would be "Prepare the meal,"but this wouldn't be too helpful in figuring out what had to be done. A two-level WBS would include:

Preparation
- Boil soup
- Boil rice
- Boil peas
- Brown chicken
- Prepare sauce
- Bake chicken, rice, and sauce
- Open wine and let it breathe

Eating
- Eat soup
- Eat entrée

This two-level WBS provides more specific guidance but still leaves a lot up to the chef (which is OK in many cases).

A more elaborate approach to preparing a WBS is to use Post-It notes to sort out the sequences. There are several possible sequences for the activities for this WBS, depending on your interpretation of the proper order in which to prepare this meal. One possible WBS showing precedence relationships (technically this is now a *precedence network*) is shown in Figure 6.1.

In this WBS only the activity names and the resources (Pot 1, etc.) the activities use are listed. Notice that I've made decisions about placing the sauce

Figure 6.1 A Post-It Note Precedence Network.

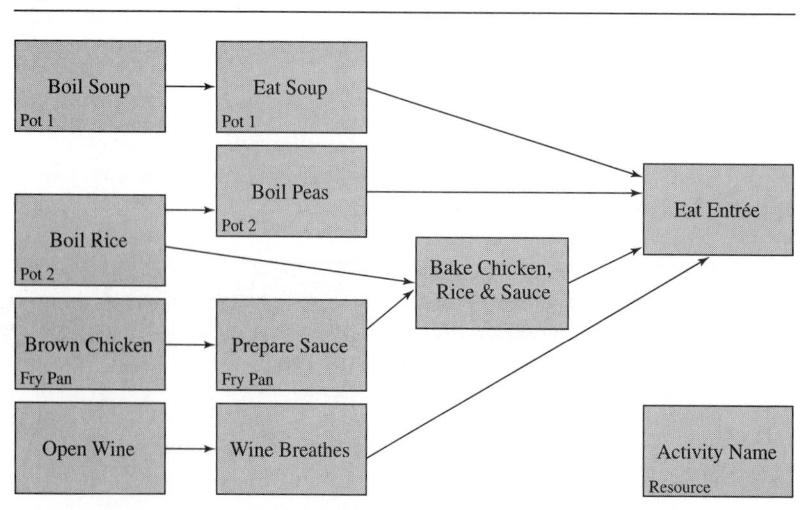

on the entrée before putting it the oven, and having the wine with the entrée rather than with the soup. You may have chosen a different sequence, perhaps to have the wine with the soup or to place the sauce on entrée after it is served. Later we'll explore how these choices affect the schedule.

REFLECTION Have you used the WBS idea for scoping projects? If not, are there places in your personal and professional life where you can immediately apply the WBS idea? How about an engineering course or design project you're working on? If you want more practice, try the office-remodeling project example at the end of the chapter.

Critical Path Method

Now that we have a precedence network for the meal-planning project, we can determine the minimum time to complete it. To do this, we go through Figure 6.1, number each node and list the time it takes (see Figure 6.2). Examine the precedence network in Figure 6.2 to determine the minimum time to complete the project. Sum the individual activity durations along each path; for example, path 1+5+10 is 35+15+25=75 min. Which path is longest?

Provided that the number of activities is not too large, problems of this type can often be solved by hand. By sketching the relationships between the individual tasks, and taking into account the amount of time each requires for completion, we can determine the total amount of time needed to get the whole process completed.

When the number of tasks gets large—say, over 20—then it's quite challenging to keep track of everything by hand. A simple and systematic way of doing this is provided by the critical path method (CPM). This method represents the flow of tasks in the form of a network. To use it, we simply have to know the duration of each of the activities, and the predecessors of each—that is, the set of activities that must have terminated before a given activity can begin.

Figure 6.2 A Precedence Network.

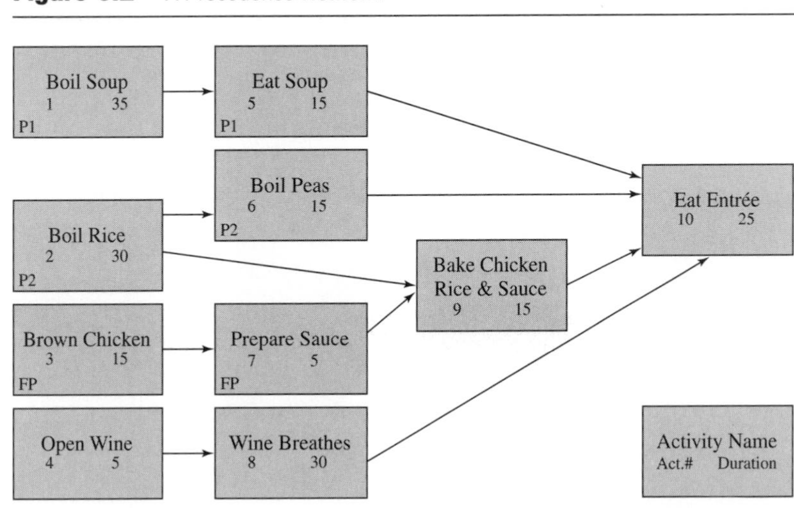

Forward Pass—Early Start (ES) and Early Finish (EF)

The first step in the procedure is to run through the network from beginning to end and mark the earliest time at which each activity can start—the early start (ES) time. In Figure 6.3, this is time is in the upper left-hand corner of each activity. This is clearly obtained by adding the earliest start of its latest starting predecessor to that predecessor's duration. When two or more activities must be completed before the next one can start (such as Boil Rice and Brown Chicken and Prepare Sauce before Bake Entrée), then the maximum must be used. The early finish (EF) time is determined by summing the early start (ES) and the duration (see Figure 6.3).

Backward Pass—Late Start (LS) and Late Finish (LF)

Similarly, a backward pass is made, establishing the latest possible starting time (late start, LS) that an activity can have, which is the latest start of the earliest starting successor, less the duration of the activity under consideration (see Figure 6.4). The result is the late finish (LF) time.

Critical Path

Activities for which the earliest and latest times turn out to be equal are called *critical.* That is, these activities cannot be delayed without increasing the duration of the entire project. The path that these activities lie on in the network is known as the *critical path.* The remaining *noncritical* activities have some "float" (sometimes referred to as slack) and can have their durations increased by some amount before they would become critical and delay the completion of the entire project.

Figure 6.3 CPM: Forward Pass—Early Times.

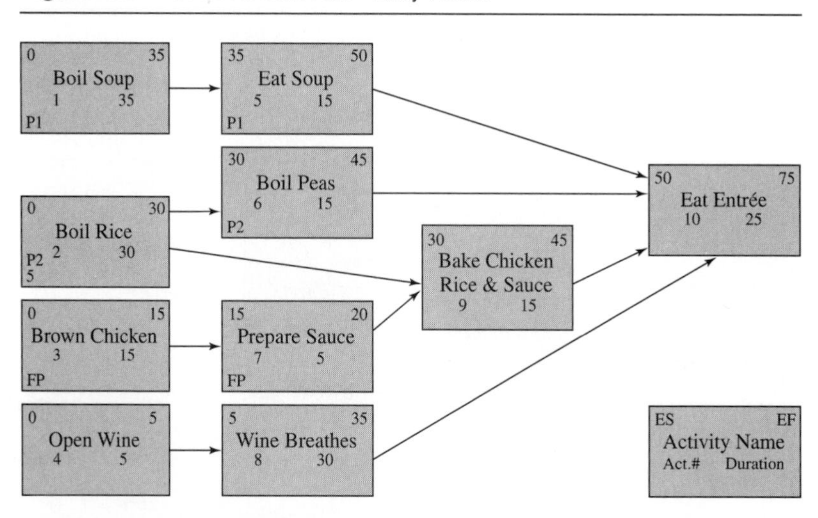

Figure 6.4 CPM: Backward Pass—Late Times.

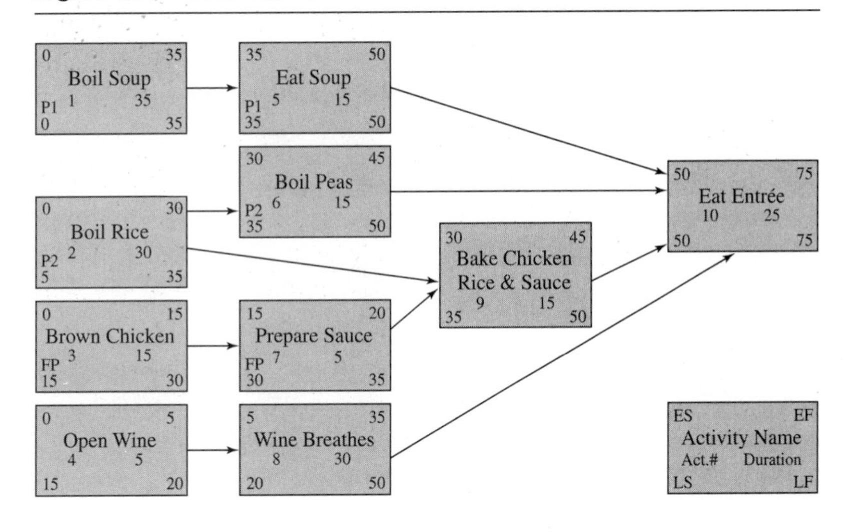

Floats

The amount by which termination of a noncritical activity can be delayed before it causes one of its successors to be delayed is called the *free float* of that activity. Technically, the free float (FF) is based on early start (ES) times, and for any activity i, the free float is equal to the minimum early start for activities following activity i minus the early start for i minus the duration (D) for i. Algebraically, the free float (FF) is determined as follows:

$$FF_i = (ES_{i+1})_{min} - ES_i - D_i$$

The amount of slack an activity has before it would cause the path on which it lies to become critical is called the *total float.* The total float of an activity is the minimum (out of all of the paths on which it lies) of the sum of its free float and those of all activities ahead of it on the path. Thus an activity is critical if its total float is zero. Technically, the total float (TF) is the difference between the late times and the early times—late start minus early start or late finish minus early finish. Algebraically, the total float (TF) for an activity i is determined as follows:

$$TF_i = LS_i - ES_i = LF_i - EF_i$$

The numerical solution to the meal planning problem is given in Table 6.1.

Another common model for representing scheduling projects is a time-scaled network, called a Gantt chart, where the activities have been laid out on a time axis. Table 6.1 and the Gantt charts in Figures 6.5 and 6.6 were prepared using the CritPath program, which is available for downloading from my website and is bundled with *How to Model It* (Starfield, Smith, and Bleloch, 1994). The CritPath program is set up to display only eight activities at a time. If you want to be able to scroll though the activities and change them, download the CritPath program from my website (www.ce.umn.edu/~smith) and play with it.

Table 6.1 Meal Planning Exercise—Critical Path Method Results

Activity	Name	Duration	Resources	Early Start	Early Finish	Late Start	Late Finish	Float Total	Float Free	Current Start	Critical Path
1	Boil soup	35	1	0	35	0	35	0	0	0	Yes
2	Boil rice	30	1	0	30	5	35	5	0	0	No
3	Brown chicken	15	1	0	15	15	30	15	0	0	No
4	Open wine	5	1	0	5	15	20	15	0	0	No
5	Eat soup	15	1	35	50	35	50	0	0	35	Yes
6	Boil peas	15	1	30	45	35	50	5	5	30	No
7	Prepare sauce	5	1	15	20	30	35	35	15	10	No
8	Wine breathes	30	1	5	35	20	50	15	15	5	No
9	Bake entrée	15	1	30	45	35	50	5	5	30	No
10	Eat entrée	25	1	50	75	50	75	0	0	50	Yes

Figure 6.5 Gantt Chart, Activities 1–8.

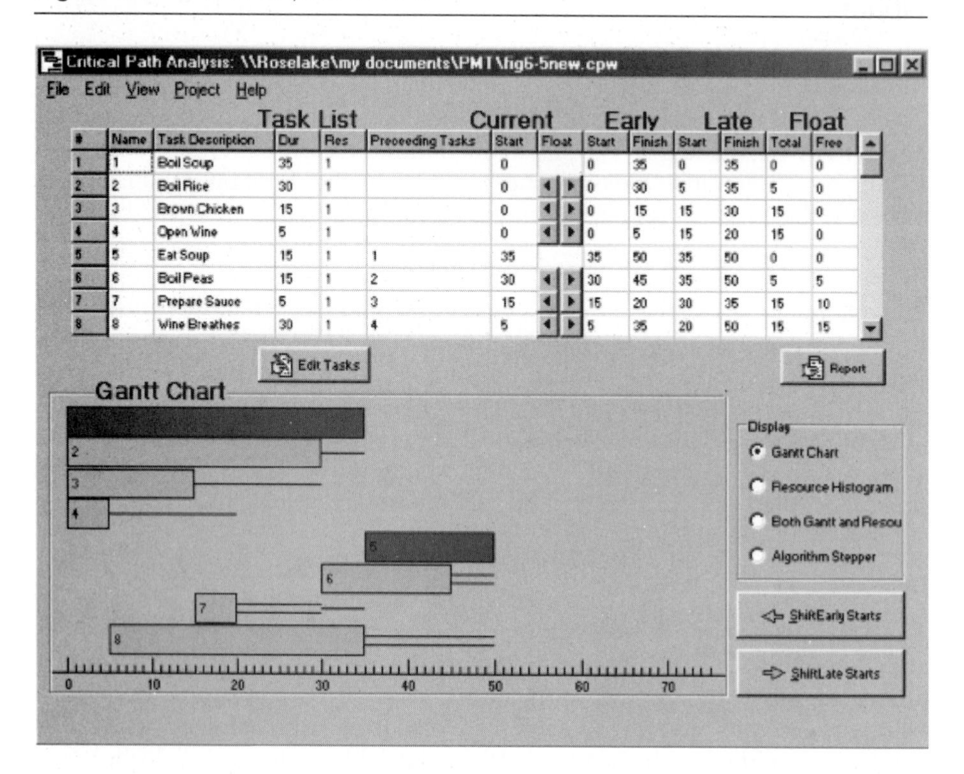

If you are still having difficulty understanding the differences between free float and total float or are struggling with the critical path calculations, use the algorithm stepper in the CritPath program. It will walk you through the process using a graphical representation (see Figure 6.7).

Figure 6.6 Gantt Chart, Activities 3–10.

Critical Path Analysis: \\Roselake\my documents\PMT\fig6-5new.cpw

File Edit View Project Help

	Task List					Current		Early		Late		Float	
#	Name	Task Description	Dur	Res	Preceeding Tasks	Start	Float	Start	Finish	Start	Finish	Total	Free
3	3	Brown Chicken	15	1		0		0	15	15	30	15	0
4	4	Open Wine	5	1		0		0	5	15	20	15	0
5	5	Eat Soup	15	1	1	35		35	50	35	50	0	0
6	6	Boil Peas	15	1	2	30		30	45	35	50	5	5
7	7	Prepare Sauce	5	1	3	15		15	20	30	35	15	10
8	8	Wine Breathes	30	1	4	5		5	35	20	50	15	15
9	9	Bake Entree	15	1	27	30		30	45	35	50	5	5
10	10	Eat Entree	25	1	5698	50		50	75	50	75	0	0

Edit Tasks   Report

Gantt Chart

3
4
5
6
7
8
9
10
0 10 20 30 40 50 60 70

Display
Gantt Chart
Resource Histogram
Both Gantt and Resou
Algorithm Stepper

ShiftEarly Starts
ShiftLate Starts

Notice that there is no gap in the path that includes activities Open Wine (1), Eat Soup (5), and Eat Entrée (10). That means, of course, that they are on the critical path. Also notice how there is a gap after activities Wine Breathes (8), Prepare Sauce (7), Bake Entrée (9) and Boil Peas (6), which means they have free float in addition to having total float. The activities Brown Chicken (3) and Open Wine (4) are followed by a gap farther down the path, but not by an immediate gap; this means that they have total float but not free float.

> INDIVIDUAL REFLECTION Take a few minutes to think about the advantages and disadvantages of the two representations of the meal-planning project—the precedence network (Figure 6.4) and the Gantt chart (Figures 6.5 and 6.6). What are the unique features of each? What specific features does each represent? Where is each appropriate?

In the above Reflection, you may have concluded that both the precedence network and the Gantt chart are essential for understanding complex projects and communicating project information. The Gantt chart is a time-scaled network, because it represents time directly. It gives a clear picture of the duration of events, but it doesn't directly show the nature of the interdependence—that is, what has to follow what. The precedence network, on the other hand, clearly shows the interdependence—precedence and simultaneity—that is, what has to follow what and what can be going on at the same time, but it doesn't directly show the time required for each activity.

Figure 6.7 Critical Path Algorithm Stepper.

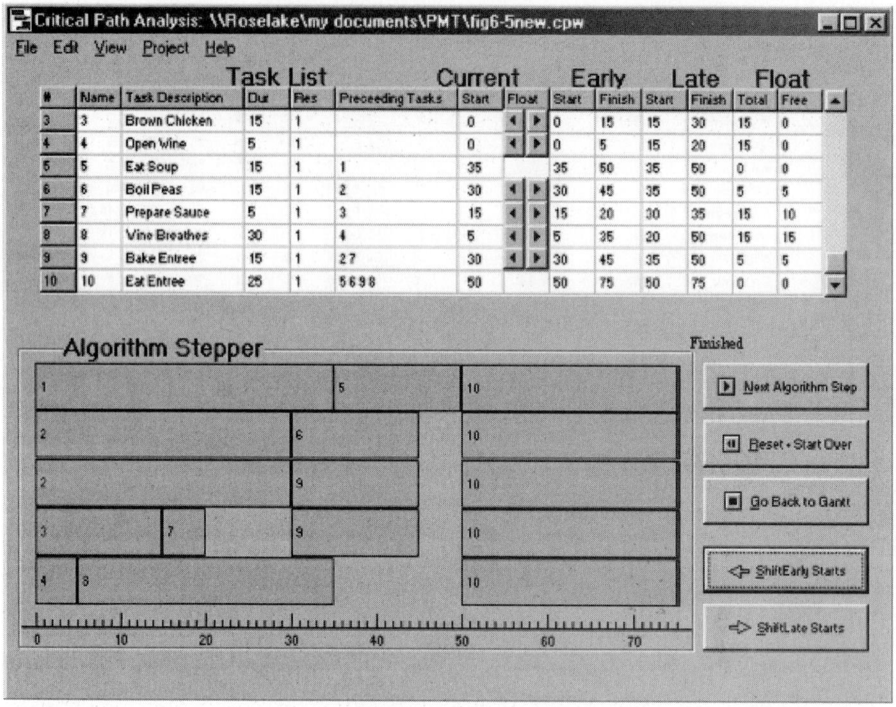

The CritPath program and most project-scheduling software uses the precedence network representation to do the critical path calculations. Many people find the Gantt chart most useful for tracking project progress.

Critical Path Method Summary

In summary, the sequence of steps to apply the critical path method to project scheduling is as follows:

1. Develop a work breakdown structure (WBS).
2. Connect the activities in the WBS by arrows that indicate the precedence.
3. Perform the critical path analysis calculations either by hand for a simple problem or using computer software.
4. Create graphical representations—precedence network and Gantt chart—that suit your purposes.

Bus Shelter Construction Example

Now that we've worked through the meal-planning exercise is some detail, let's try another example. Consider the construction project outlined in Table 6.2. Note that the precedence relationships are specified, so you don't have to create a work breakdown structure; however, developing a precedence net-

Table 6.2 Bus Shelter Construction Example

Job	Name	Duration	Resources	Predecessors
1	Shelter Slab	2	2	5
2	Shelter Walls	1	1	1
3	Shelter Roof	2	2	2,4
4	Roof Beam	3	2	2
5	Excavation	2	3	
6	Curb and Gutter	2	3	5
7	Shelter Seat	1	2	4,6
8	Paint	1	1	7
9	Signwork	1	2	2,6

work is an important step. The Resources column specifies the number of people required for each task.

Determine the minimum time required to complete the bus shelter. Develop a precedence network. Identify the critical path. Draw a Gantt chart for the project. Give it a try before reading further. (A precedence network is sketched out in Figure 6.8.)

Next, perform the critical path analysis calculations. You may do this by hand if you want more practice, or you can use CritPath or another commercial project-scheduling program. The table and Gantt chart for the first eight activities are shown in Figure 6.9.

Look carefully at the critical path method results and the Gantt chart. Notice that there is more than one critical path. The presence of multiple critical paths can be seen on the Algorithm Stepper, because it displays all the paths through the network. Get the CritPath software and try it. Table 6.3 shows the entire set of results for the bus shelter construction project.

Try developing and setting up your next project using the critical path method. Do the calculations by hand a couple of times to familiarize yourself with the forward pass and backward pass of the algorithm. Then use CritPath or another commercial project-scheduling program.

Figure 6.8 Precedence Network for Bus Shelter Construction.

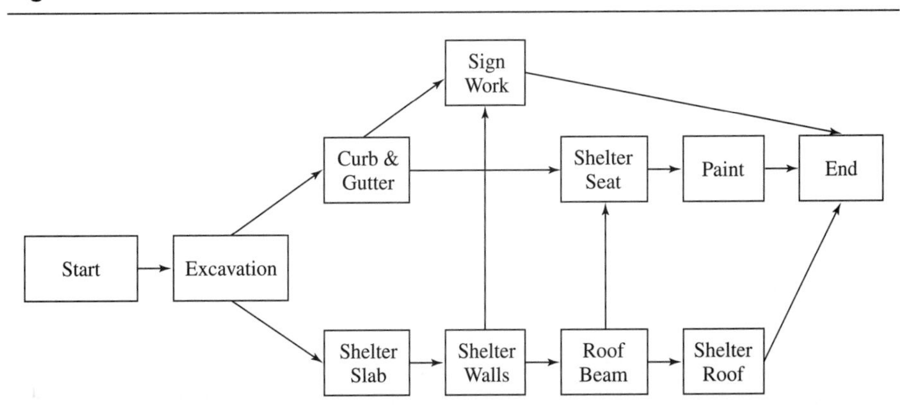

Figure 6.9 Gantt Chart for Bus Shelter Construction.

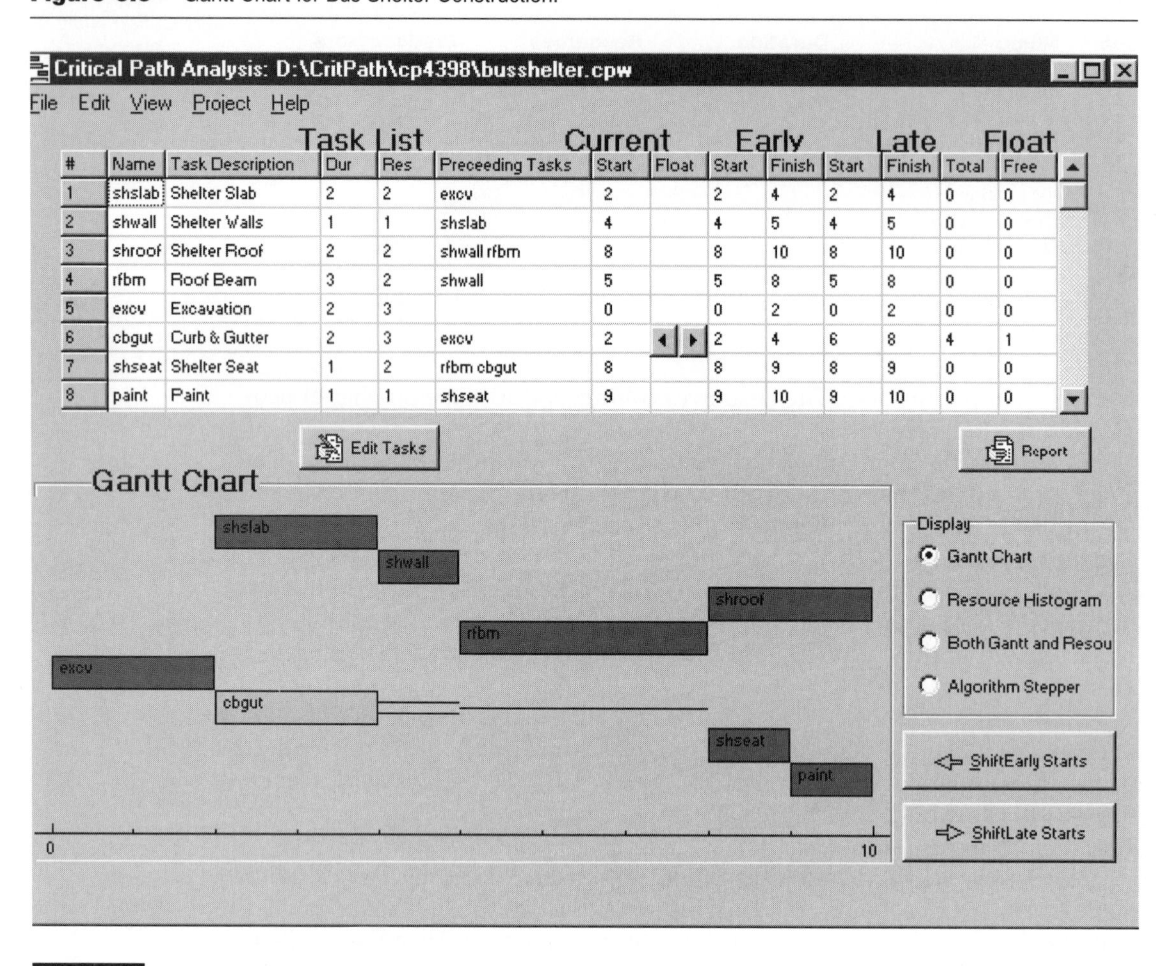

Table 6.3 Bus Shelter Construction—Critical Path Method Results

Activity	Name	Duration	Resources	Early Start	Early Finish	Late Start	Late Finish	Float Total	Float Free	Current Start	Critical Path
1	Shelter Slab	2	2	2	4	2	4	0	0	2	Yes
2	Shelter Walls	1	1	4	5	4	5	0	0	4	Yes
3	Shelter Roof	2	2	8	10	8	10	0	0	8	Yes
4	Roof Beam	3	2	5	8	5	8	0	0	5	Yes
5	Excavation	2	3	0	2	0	2	0	0	0	Yes
6	Curb & Gutter	2	3	2	4	6	8	4	1	2	No
7	Shelter Seat	1	2	8	9	8	9	0	0	8	Yes
8	Paint	1	1	9	10	9	10	0	0	9	Yes
9	Signwork	1	2	5	6	9	10	4	4	5	No

Resource Leveling

Critical activities, having no slack, cannot be extended or shifted without upsetting the scheduled completion of the project. However, the slack afforded by noncritical activities can be exploited to provide the best distribution of resources over the duration of the entire project. For example it might be difficult or expensive to hire more than a certain number of programmers or tradespeople at any one time. By shifting noncritical activities within their floats, it is possible to spread the distribution of personnel more evenly over the span of the project. At other times it may be beneficial to load the distribution in a certain way—for example, if work over a holiday period is to be minimized. Decisions of these kinds can be made only when the constraints (i.e., earliest and latest start times) of the schedule have been determined. Look at what happens to the resource histogram for the bus shelter project when you shift all activities to their early start times (Figure 6.10). What do you expect would happen if you shifted all activities to their late start times? See Figure 6.11.

Figure 6.10 Resource Histogram for Bus Shelter Construction.

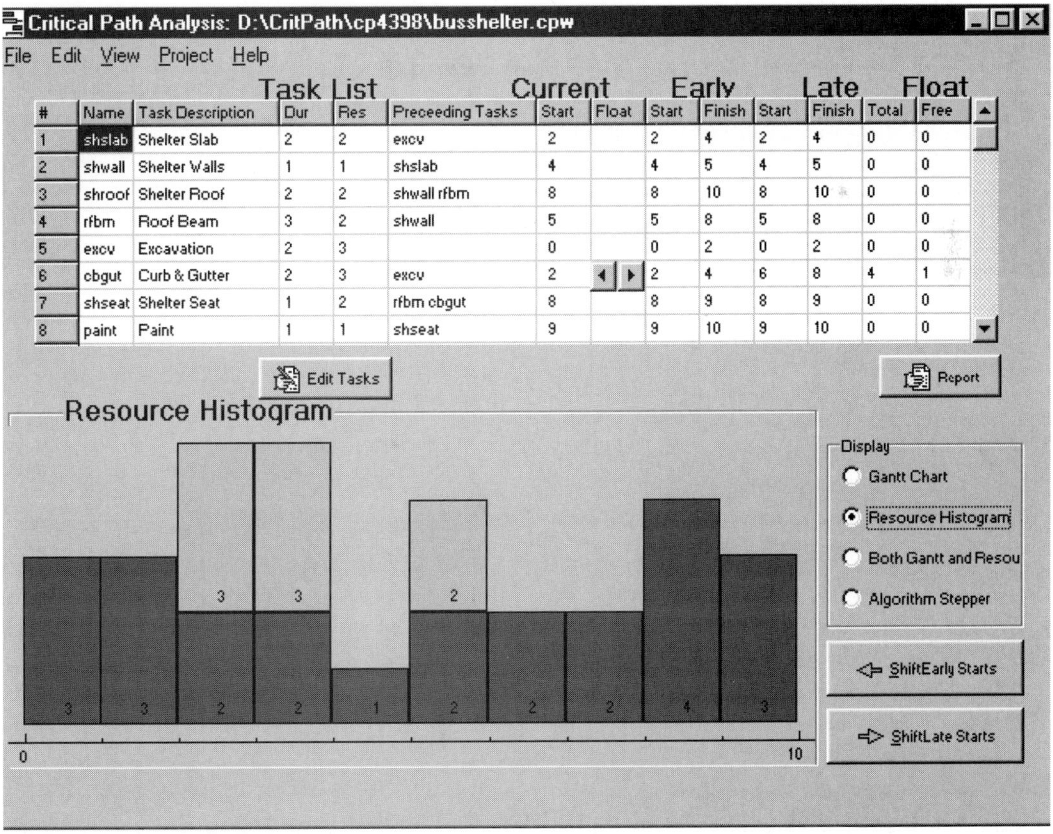

Critical Path Analysis: D:\CritPath\cp4398\busshelter.cpw

File Edit View Project Help

Task List						Current		Early		Late		Float	
#	Name	Task Description	Dur	Res	Preceeding Tasks	Start	Float	Start	Finish	Start	Finish	Total	Free
1	shslab	Shelter Slab	2	2	excv	2		2	4	2	4	0	0
2	shwall	Shelter Walls	1	1	shslab	4		4	5	4	5	0	0
3	shroof	Shelter Roof	2	2	shwall rfbm	8		8	10	8	10	0	0
4	rfbm	Roof Beam	3	2	shwall	5		5	8	5	8	0	0
5	excv	Excavation	2	3		0		0	2	0	2	0	0
6	cbgut	Curb & Gutter	2	3	excv	2		2	4	6	8	4	1
7	shseat	Shelter Seat	1	2	rfbm cbgut	8		8	9	8	9	0	0
8	paint	Paint	1	1	shseat	9		9	10	9	10	0	0

Edit Tasks

Report

Resource Histogram

3 3 3 3 2

3 3 2 2 1 2 2 2 4 3

0 10

Display

Gantt Chart

Resource Histogram

Both Gantt and Resou

Algorithm Stepper

ShiftEarly Starts

ShiftLate Starts

Figure 6.11 Resource Histogram—Latest Times.

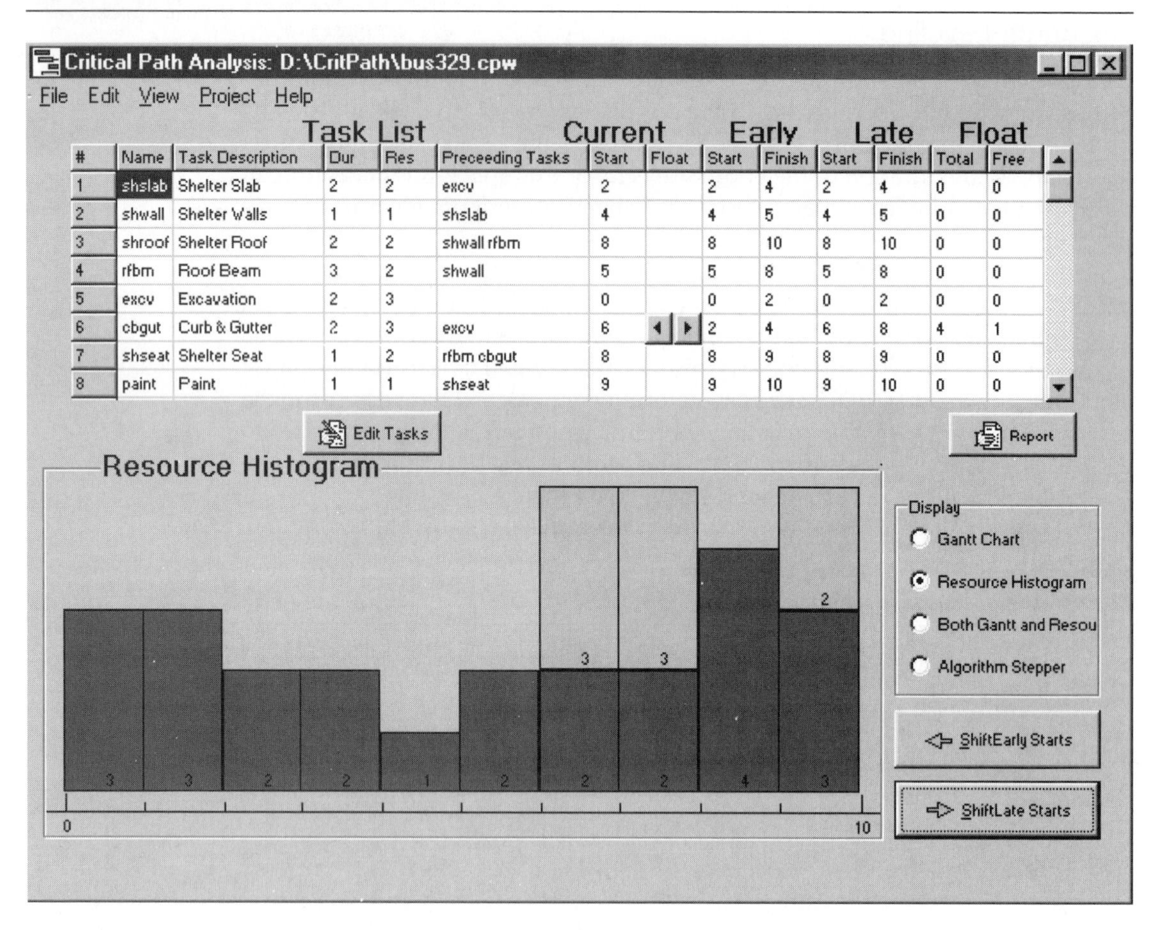

#	Name	Task Description	Dur	Res	Preceeding Tasks	Current Start	Current Float	Early Start	Early Finish	Late Start	Late Finish	Float Total	Float Free
1	shslab	Shelter Slab	2	2	excv	2		2	4	2	4	0	0
2	shwall	Shelter Walls	1	1	shslab	4		4	5	4	5	0	0
3	shroof	Shelter Roof	2	2	shwall rfbm	8		8	10	8	10	0	0
4	rfbm	Roof Beam	3	2	shwall	5		5	8	5	8	0	0
5	excv	Excavation	2	3		0		0	2	0	2	0	0
6	cbgut	Curb & Gutter	2	3	excv	6		2	4	6	8	4	1
7	shseat	Shelter Seat	1	2	rfbm cbgut	8		8	9	8	9	0	0
8	paint	Paint	1	1	shseat	9		9	10	9	10	0	0

Compare Figures 6.10 and 6.11 to help decide what you think would be the best allocation of resources. Notice that there is not too much that you can do to level the resources in this case, which is sometimes the situation with real projects.

Cost Considerations

On some projects you might work through the critical path calculations and find that the required project duration is greater than the time you have available. Or, more commonly, you might get behind during the project due to weather, late deliveries, work delays, and so on. In such cases, you could, of course, go to your supervisor or professor and ask for an extension.

Alternatively, you could add resources (e.g., people, overtime) to activities on the critical path to decrease their duration, thus decreasing the time for the entire project. Technically, this is known as "crashing" a project. Why wouldn't we add resources to noncritical activities? Let's work through the example shown in Table 6.4 to get a better sense of how this works.

Table 6.4 CPM/Cost Example

Task	Precedence	Normal Time	Crash Time	Normal Cost	Incremental Cost/Day
A	–	8	4	800	300
B	–	4	3	600	100
C	B	2	1	1000	400
D	A, C	3	2	300	200

Figure 6.12 shows the "normal" schedule duration. The normal schedule cost is just the sum of the normal costs for the four activities—$2700.

As the project manager, you would have to choose which activities along the critical path to add resources to, in order to decrease their duration. What criteria would you use to choose the activities? Would you crash activities early in the schedule first? The cost-conscious project manager would crash the activities that had the minimum cost per unit of time saved. Often convenience, availability, and other factors must be considered.

Figure 6.12 Gantt Chart for Cost Example.

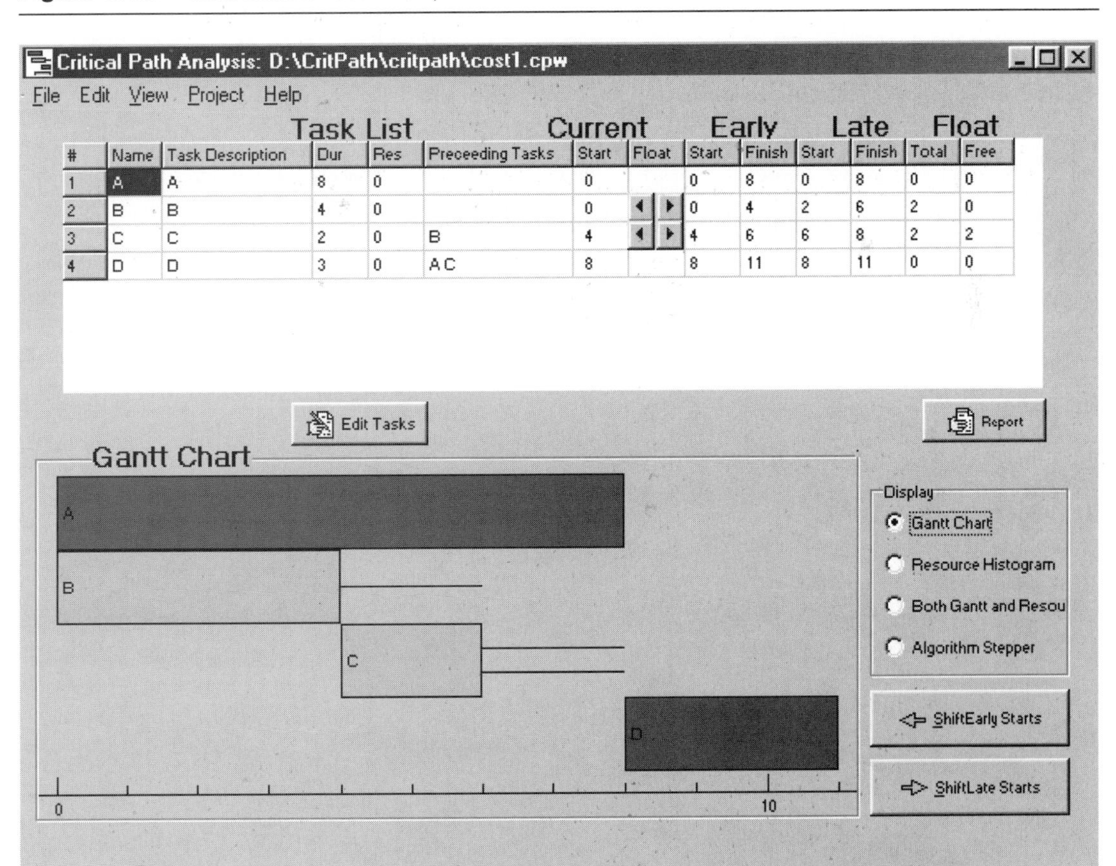

Let's look at what happens as we decrease the duration of activities on the critical path for the example above (see Figure 6.12). The lowest-cost choice is Activity D. By decreasing the duration of Activity D from 3 to 2, the overall project duration decreases to 10 and the cost increases to \$2700 + \$200 = \$2900. Next we can add resources to Activity A to decrease it to 7. The overall project duration becomes 9 and the cost increases to \$3000. Adding more resources to A to decrease it to 6 decreases the overall project duration to 8 and increases the cost to \$3300. The updated Gantt Chart is shown in Figure 6.13. We can continue to decrease the duration of the overall project, but now we must add resources to more than one activity and hence the cost increases at a higher rate.

As you add resources to critical activities and decrease the duration along the critical path, eventually more and more activities become critical.

INDIVIDUAL AND GROUP REFLECTION What are some good strategies for using the float over the life of a project? Do you let things slide early on, thus using the float up early? Do you wait until later in the project to use the float? As a project manager, how do you recommend that the float be utilized?

Figure 6.13 Gantt Chart—All Paths Critical.

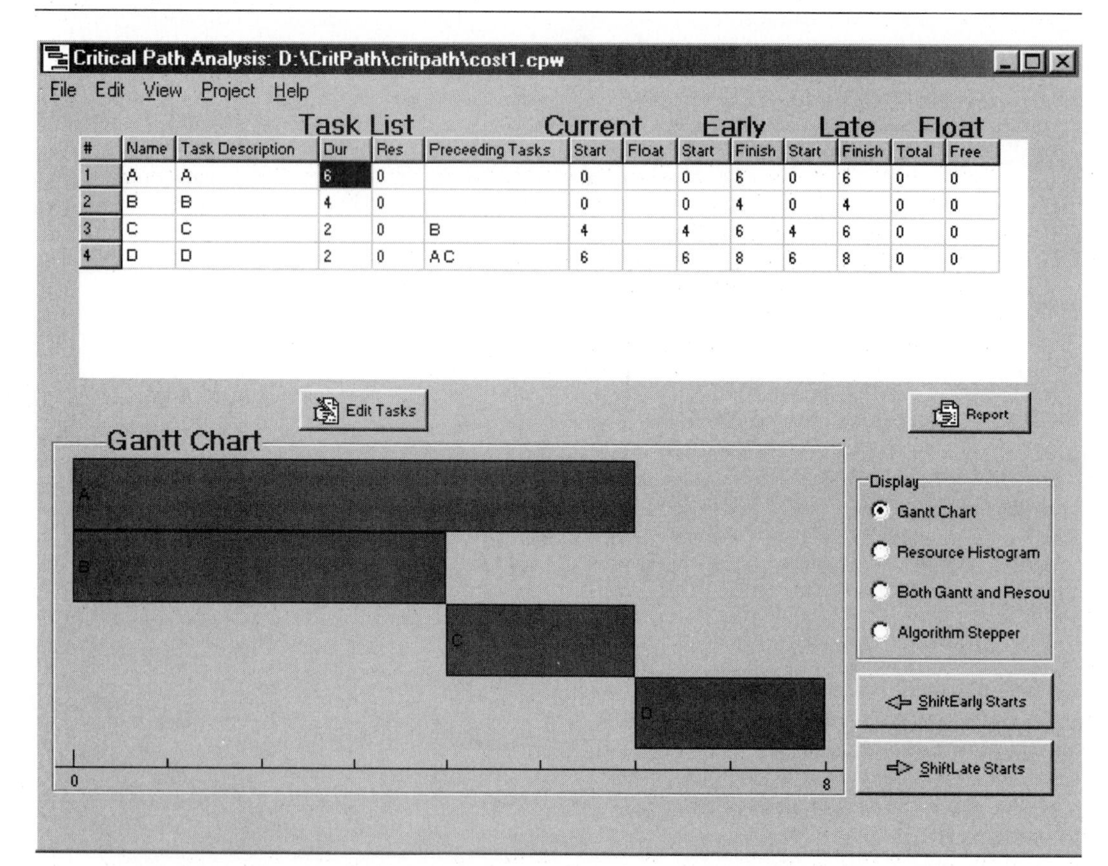

The Role of Computer-Based Project Management Software

It is not hard to see that if we were to add a few more activities, the problem would soon become unmanageable by hand. Further, if changes have to be made to either the order in which activities must occur or the time in which they can be completed, the entire process would have to be repeated. The advantage of the critical path method is that it is indeed systematic. It can be described as a formal set of instructions that can be followed by a computer. Alterations in the data can be made repeatedly and the problem quickly solved, again by the machine. This would enable us to obtain the benefit of what-if analysis, the process of making changes and seeing effects of those changes immediately. Such analysis gives the user an intuitive feel for the problem. The role of the computer will be addressed in more detail in Chapter 9.

Reflection: Avoiding Analysis Paralysis

I've learned that engineering students love the stuff in this chapter; it resonates with them. I've also seen students get so involved in the analysis that they forget what they're analyzing, and why they're doing it. Analysis is important, but in engineering, analysis must be purposeful—it must be focused on developing understanding so that a decision can be made. Also, it's essential to decide how good is good enough. This is a really tough decision when we know that if we spend more time and resources we'll get a better answer.

Questions

1. What is a work breakdown structure (WBS)? Why is it important? What are some of the types of WBS?
2. What is the critical path method (CPM)? What are free float and total float?
3. Explain the differences between resource leveling and crashing a project.
4. How have you typically managed complex projects in the past? How well did the approach work? Not all projects merit taking the time to develop a schedule because they can be managed by a list or by using your day planner. Where could you apply the critical path method and how do you expect it might help?
5. How does the CritPath program compare with other software packages you've used (spreadsheet, for example)?
6. What are some of the advantages and disadvantages of relying on project-scheduling algorithms?

Exercises

1. **Work Breakdown Structure Example—Office Remodeling Project**

The following activities must be accomplished to complete an office remodeling project:

Activity	Estimated Duration (Days)
Procure paint	2
Procure new carpet	5
Procure new furniture	7
Remove old furniture	1
Remove old carpet	1
Scrub walls	1
Paint walls	2
Install new carpet	1
Move in new furniture	1

Part 1: Work Breakdown Structure (WBS)

1. Create a possible work breakdown structure (WBS) for the remodeling project.

Part 2: Scheduling

1. When can the new furniture be moved in?
2. What is the minimum project duration?
3. Which activities do you have to play close attention to if you want to finish at the earliest possible time?

2. Resource Leveling

Given the following set of project data, determine the smoothest distribution of resources. Assume that resources are transferrable.

Task	Duration	Resources	Predecessor
1	7	8	
2	5	6	
3	4	4	2
4	2	4	1,3
5	3	6	2
6	1	6	5

3. Crashing

Given the following project data, determine the normal schedule cost.

Crash the project as far as possible. List the project duration and cost for each step along the way.

Task	Precedence	Normal Time	Crash Time	Normal Cost	Incremental Cost Per Day
A	–	5	3	500	300
B	A	4	2	600	100
C	A	6	4	1000	400
D	B, C	3	2	300	200

4. Scheduling Case Study

Develop a work breakdown structure, precedence network, and Gantt chart for a project you're involved with. Complete the critical path analysis calculations. Use these representations to guide the project and to review progress.

5. A Minnesota Dream Project—Lakeside Cabin Cost Estimate

Imagine you've just inherited a lakeside lot in northern Minnesota (beautiful in the summer) that has an access road, electricity, water, and sewage disposal. You've also inherited a modest sum with which to buy building materials. If you're having trouble imagining yourself being so lucky, suppose that a friend or neighbor has asked you for an estimate on how much it would cost to build a cabin and for help in designing and building. Engineers are often expected to know such things. How much would it cost for a modest, say 24-foot-square, cabin? How long would it take you to build it?

The cabin cost estimate project is a favorite of students in my project management classes. We haven't done much here with cost estimating (this is usually done in engineering economics courses), so I'll provide a little guidance. I suggest you take the following steps:

1. Guess. Take a wild guess at what it would cost for materials for a modest cabin.
2. Look up the square-foot costs for typical residential construction ($50–70 per square foot) and adjust it down for a modest cabin. Many students use $30 per square foot for a modest cabin.
3. Use a unit cost approach:
    1. Develop a floor plan.
    2. Create a detailed list of materials with associated quantities.
    3. Find the cost of the materials in a local building center brochure (or cost manual).
    4. Use a spreadsheet to list and calculate the total cost of the materials.
4. Compare the three cost estimates—guess, cost per square foot, and unit cost.

Use the procedures outlined in this chapter to determine how long it would take you to build it; that is, develop a WBS and a schedule using the critical path method.

Reference

Starfield, A. M., K. A. Smith, and A. L. Bleloch. 1994. *How to model it: Problem solving for the computer age.* Edina, MN: Burgess. (Includes disk with Critpath and WinExp.)

CHAPTER 7

Project Monitoring and Evaluation

Two important parts of project management are keeping on top of projects through monitoring and conducting end-of-the-project review through evaluation. There are many ways to keep informed, keep others informed, and coordinate projects. Informal one-on-one meetings, e-mail, and phone conversations are the most common. Other simple forms include fax, voice messages, and handwritten notes. Emerging forms of communication include Net meetings, instant messaging (Internet, PDA, and phone), and Web-based project discussion systems, such as eProject. More involved forms of information exchange are groupware (such as Lotus Notes), enterprise systems (such as Microsoft Project 2002), informal reports, formal memos and meetings, and formal presentations. Each has advantages and disadvantages.

> REFLECTION What is your experience with face-to-face, phone, fax, e-mail, and handwritten communications? Under what conditions do you prefer each? What is your experience with Web-based information exchange–Net meetings, instant messaging, Web discussions. How well do these work for you?

Tom Peters (1989) popularized the idea of "management by wandering around" (MBWA), which has become an important aspect of most project managers' daily lives. Effective project managers stay in touch. More recently, Peters (1999) has said that "all work is project work" and that individuals are identified and recognized by their project work.

This chapter begins by discussing meetings, progresses through successively more complex ways to monitor the work of projects and teams, and ends with strategies for evaluating projects.

Meetings

Although meetings are a principal way of keeping up-to-date, they are also one of the banes of project managers and many others. Dressler (cited in Lewis, 1996) lists some of the major complaints people have about meetings:

- Their purpose is unclear.
- Participants are unprepared.

- Key people are absent or missing.
- The conversation veers off track.
- Participants don't discuss issues but instead dominate, argue, or take no part at all.
- Decisions made during the meeting are not followed up on.

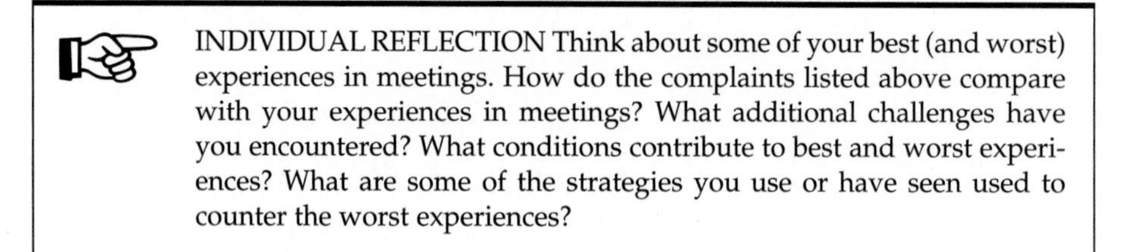

INDIVIDUAL REFLECTION Think about some of your best (and worst) experiences in meetings. How do the complaints listed above compare with your experiences in meetings? What additional challenges have you encountered? What conditions contribute to best and worst experiences? What are some of the strategies you use or have seen used to counter the worst experiences?

Five guidelines from the book *Meetings: Do's, Don'ts, and Donuts* (Lippincott, 1994) can help alleviate some of the most common problems:

1. State in a couple of sentences exactly what you want your meeting to accomplish.
2. If you think a meeting is the best way to accomplish this, then distribute an agenda to participants at least 2 days in advance.
3. Set ground rules to maintain focus, respect, and order during the meeting.
4. Take responsibility for the meeting's outcome.
5. If your meeting isn't working, try other tools, such as brainstorming.

Using a meeting process, such as the one outlined in *The Team Handbook* (Scholtes, Joiner, and Streibel, 1996) also can help. The authors describe the three-part process as follows:

Before

Plan.
Clarify meeting purpose and outcome.
Identify meeting participants.
Select methods to meet purpose.
Develop and distribute agenda.
Set up room.

During

Start: check-in, review agenda, set or review ground rules, clarify roles.
Conduct: cover one item at a time, manage discussions, maintain focus and pace.
Close: summarize decisions, review action items, solicit agenda items for next meeting, review time and place for next meeting, evaluate the meeting, thank participants.

After

Follow-up: distribute or post meeting notes promptly; file agendas, notes, and other documents; do assignments.

The box "How to Run a Meeting" also provides good advice for running effective meetings.

How to Run a Meeting

PLAN THE MEETING

- Be clear on objectives of the meeting
- Be clear WHY you need the meeting
- List the topics to be addressed

INFORM

- Make sure everyone knows exactly what is being discussed, why and what you want from the discussion
- Anticipate what people and information may be needed and make sure they are there

PREPARE

- Prepare the logical sequence of items
- Allocate time on the basis of importance, not its urgency

STRUCTURE AND CONTROL

- Put all evidence before interpretation and action
- Stop people from jumping back and going over old ground

SUMMARIZE AND RECALL

- Summarize all decisions and record them with the name of the person responsible for any action

Source: Adapted from "Meetings, Bloody Meetings" starring John Cleese, 1993.

One emerging idea in this area is that of virtual meetings. John Leeper, Information Services director at Ryan Construction, reported recently that Ryan Construction is making extensive use of Webex for meetings (Leeper, 2002).

Monitoring Team Effectiveness

An important part of any team meeting is taking time to reflect on how well the team worked together. One common method for monitoring the effectiveness of group work is the plus/delta team processing approach typically attributed to Boeing. Near the end of the meeting the team stops working on the task and spends a few minutes discussing how well they worked. The team makes a list that records what went well on one side (+) and what they could do even better on the other (Δ). Other methods include individual reflection using instruments such as the one in Figure 7.1. Team members each fill in the chart and then discuss each other's scores and comments.

More complex monitoring involves collecting data on individual participation in the team. Several observation forms are available. The one I start students with in my project management classes is shown in Figure 7.2. Any of the task and maintenance behaviors (see Chapter 3) may be listed in the rows. Group members take turns observing the group and recording each member's participation. They then provide feedback about the team's functioning during the processing phase. A rule of thumb that I commonly use is "Keep the feedback specific, descriptive, immediate, and positive." I give negative feedback only if the person requests it. If negative feedback is requested, the person is usually ready to hear it. Only then will it be helpful. More rules of thumb for feedback include the following, from Scholtes (1988):

Figure 7.1 A Sample Individual Reflection Instrument.

For each trait, rate the team on a scale of 1 to 5:
1 = Not present (opposite trait present)
2 = Very poor (not much evidence of positive trait)
3 = Poor (some positive trait seen)
4 = Good (positive trait evident more than opposite trait)
5 = Very good (large amounts of evidence of positive trait)

Positive Trait	Score	Comments
Safety		
Inclusion		
Free interaction		
Appropriate level of interdependence		
Cohesiveness		
Trust		
Conflict resolution		
Influence		
Accomplishment		
Growth		

Source: Uhlfedler, 1997.

Guidelines for Constructive Feedback

- Acknowledge the need for feedback.
- Give positive feedback (give negative feedback only if the recipient asks for it).
- Understand the context.
- Know when to give feedback.
- Know how to give feedback.
- Be descriptive.
- Don't use labels.
- Don't exaggerate.
- Don't be judgmental.
- Speak for yourself.
- Talk first about yourself, not about the other person.
- Phrase the issue as a statement, not a question.
- Restrict your feedback to things you observed.
- Help people hear and accept your compliments when giving positive feedback.

The idea of observing teams in action and trying to document their interactions and effectiveness is a common research approach (Bucciarelli, 1996; Cuff, 1991; Donnellon 1996; Minneman, 1991; Seely, Brown, and Duguid, 1991; Wenger and Snyder, 2000). It is also common in business and industry. A form that I've found very useful is shown in Figure 7.3. This was developed by

Figure 7.2 Observation Sheet.

Observation Category	**Names**	**Total**
Task Contributes ideas		
Maintenance Encourages		
Total		

Notes:

Observation Directions

- Move your chair so you can see each member of the group clearly without interacting with them.
- Write the name of each person in the group at the top of one of the columns on the Observation Sheet.
- Watch each person systematically. As you see each person display one of the two behaviors specified (contributes ideas or describes feelings) place a hatch mark below his/her name, in the box to the right of the appropriate behavior.
- *Task* means anything that helps the group accomplish its task. For example, *Contributes ideas* means giving an idea related to one of the questions on the worksheet and/or to something said by another group member related to the task.
- *Maintenance* means anything that helps improve working relationships in the group. *Encourages* means praising other's ideas or inviting others to contribute.
- You may make some notations below the grid that may help you explain some scoring.
- When time is called, total the hatch marks in each column and across each row.
- When you give feedback to the group you will only give them the column totals and the row totals. Let them see the sheet. DO NOT INTERPRET WHAT YOU SAW, OR WHAT THE TOTALS MIGHT SUGGEST. You may be tempted to soften what you say. Don't allow this to happen. You are not criticizing, you are only reporting what you saw, related to very specific behaviors.
- It is the job of the group to discuss what the totals might suggest about how they functioned as a group, and to develop one or two sentences that capture this thought.

Xerox and is included in their interactive skills workbook (Xerox Corporation, 1986). The second part of the figure provides definitions and examples of behavior categories in the worksheet.

The purpose of collecting data on team functioning by observing and other means is not only to provide data for monitoring but also to help each member learn how to attend to how the team is performing. Even if you don't have an opportunity to systematically observe a team, read the definitions and examples in Figure 7.3 and think about how you can expand your repertoire of behaviors.

Figure 7.3 Xerox's Interactive Skills Coding Worksheet.

Behavior						Totals	%
Initiating							
Proposing							
Building							
Reacting							
Supporting							
Disagreeing							
Defending/attacking							
Clarifying							
Testing understanding							
Summarizing							
Seeking information							
Giving information							
Totals							
Process							
Shutting out							
Bringing in							

Definitions and Examples

Behavior Category	Definition	Examples
Proposing	A behavior that puts forward a new suggestion, proposal, or course of action.	"Let's deal with that one tomorrow." "I suggest we add more resources to . . ."
Building	A behavior that usually takes the form of a proposal, but that actually extends or develops a proposal made by another person.	". . . and your plan would be even better if we added a second reporting stage." "If I can take that further, we could also use the system to give us better cost control."
Supporting	A behavior that makes a conscious and direct declaration of agreement with or support for another person, or for their concepts and opinions.	"Yes, I go along with that." "Sounds OK to me." "Fine. I accept that."
Disagreeing	A behavior that states a direct disagreement or that raises obstacles and objections to another person's concepts or opinions. Disagreeing is about issues.	"No, I don't agree with that." "Your third point just isn't true." "What you're suggesting won't work."
Defending/attacking	A behavior that attacks another person, either directly or by defensiveness. Defending/attacking usually involves value judgments and often contains emotional overtones. It is usually about *people,* not issues.	"That's stupid." ". . . and your third point is either stupid or an out-and-out lie." "Don't blame me, it's not *my* fault; it's *John's* responsibility."

Figure 7.3 Xerox's Interactive Skills Coding Worksheet. (Continued)

Behavior Category	Definition	Examples
Testing understanding	A behavior that seeks to establish whether or not an earlier contribution has been understood. It differs from seeking information in that it is an attempt to ensure agreement or consensus of some kind, and refers to a prior question or issue.	"Can I just check to be sure we're talking about the same thing here?" "Can I take it that we all now agree on this?"
Summarizing	A behavior that summarizes or restates, in a compact form, the content of previous discussions or events.	"So far, we have agreed (a) To divide responsibilities, (b) To meet weekly, (c) To finish the draft proposal by . . ."
Seeking information	A behavior that seeks facts, opinions, or clarification from another person.	"What timeline did we agree to?" "Can anyone tell me what page that table is on?" "Have you checked that thoroughly?"
Giving information	A behavior that offers facts, opinions, or clarification to other people.	"I remember a case like that last year." "There are at least three alternatives."
Bringing in*	A behavior that invites views or opinions from a member of the group who is not actively participating in the discussion.	"Dick, have you anything to say on this one?" "Cheryl has been very quiet. I wonder whether she has anything she would like to say here."
Shutting out*	A behavior that excludes another person or persons, or reduces their opportunity to contribute. Interrupting is the most common form of shutting out.	Jose: "What do you think, Bob?" Karl: "What I think is. . ." (Karl is shutting out Bob)

*Characteristic of a process behavior

Team Talk Analysis

An even more sophisticated approach to processing the work of teams was developed by Donnellon (1996), who claims, "Not all groups are teams." Although, as I mentioned in Chapter 2, the words *group* and *team* are often used interchangeably, it is important to distinguish between the gathering of people into groups and the purposeful formation of a team. A *team,* according to Donnellon, is "a group of people who are necessary to accomplish a task that requires the continuous integration of the expertise distributed among them" (p. 10).

Donnellon studied team talk and devised six dimensions along which to assess teams: identification (with what group team members identify); interdependence (whether team members feel independent from or interdependent with one another); power differentiation (how much team members use the differences in their organizational power); social distance (whether team members feel close to or distant from one another socially); conflict management tactics (whether members use the tactics of force or collaboration to manage their conflicts); and negotiation process (whether the team uses a win–lose or a win–win process). Donnellon then used these dimensions to differentiate between nominal teams and real teams, as shown in Table 7.1.

The dimensions shown in the table are consistent with the underlying conceptual framework in this book—the importance of interdependence, low power differentiation, close social distance (trusting relationships), constructive conflict management strategies, and win–win dynamics. I encourage you to examine the groups and teams you are a member of along these dimensions. Donnellon's team talk audit for assessing team dynamics is included in her book. Use this instrument to attend to the team talk, reflect on what it tells you about the team, and then plan how you will discuss the assessment with the team.

Donnellon also described five types of teams, based on the categorization in her six dimensions. In Table 7.2, I've summarized some of the more direct paths. Think about where your team fits.

Donnellon's work indicates that there are very few paths to collaborative team profiles, a conclusion borne out in the work of Katzenbach and Smith (1993) and Bennis and Beiderman (1997), whose case studies note that very few teams perform at the highest levels. With these dimensions in mind, carefully examine your group and team experiences and then explicitly discuss the performance (functioning) of your team to help you decide (1) to leave if your

Table 7.1 Nominal Versus Real Teams

	Nominal Team	Real Team
Identification	Functional group	Team
Interdependence	Independence	Interdependence
Power differentiation	High	Low
Social distance	Distant	Close
Conflict management tactics	Forcing, accommodating, avoiding	Confronting, collaborating
Negotiation process	Win–lose	Win–win

Source: Donnellon, 1996.

Table 7.2 Team Talk Dimensions

Identification	Interdependence	Power Differentiation	Social Distance	Negotiation	Conflict Management	Profile
Team	High	Low	Close	Win–Win	Collaborative	**Collaborative**
Team	Moderate	Low	Close	Win–Win	Force–Avoid	Mostly Collaborative
Team	Moderate	High	Close			**Emergent**
Both	High	High	Distant	Win–Lose	Force–Avoid	**Adversarial**
Both	Low	High	Distant	Win–Lose	Force–Avoid	**Adversarial**
Function	Low	Low	Distant			**Nominal**
Function	Low	High	Distant	Win–Lose	Force–Avoid	**Doomed**

team is doomed, (2) refine the team if you're in the middle, or (3) celebrate and continue performing if you're a collaborative team.

Using the monitoring and processing formats described above to systematically reflect on the group's performance on both the task and their work with one another by the processing formats described above will help the group achieve its goals and help the members get better at working with one another. Group processing takes time and commitment and is typically difficult for highly motivated, task-oriented individuals. Spending a little carefully structured time on how the group is functioning can make an enormous difference in the group's effectiveness and the quality of the work environment.

Peer Assessment

Peer assessment, a process of rating other team members, often anonymously, is becoming more common in engineering teams. One of the best resources for peer assessment is *The Team Developer* (McGourty and De Meuse, 2001). The authors use an anonymous computer-based peer assessment system in which each individual rates each of the other team members in four team effectiveness dimensions—communication, decision making, collaboration, and self-management—using a 5-point scale. Each member then has access to her or his self-rating as well as the overall average for the team. McGourty and De Meuse stress the importance of using the information for development. They recommend that the individual and the team focus on using the information to help them improve and solve problems.

Peer assessment has also been used in merit review (as in merit reviews for raises and promotions). The most common system is the 360-degree evaluations that were used extensively at GE. GE abandoned the 360s because "Like anything driven by peer input, the system is capable of being 'gamed' over the long haul. People began saying nice things about one another so they would all come out with good ratings" (Welch and Byrne, 2001, p. 183). GE currently uses 360s only in special situations. Peer assessment of team members is subject to the same concern—everyone might be too nice and give others a high rating, which makes the assessment of limited value for improvement. Effective development results from a climate that promotes improvement. Peer assessment might help create such a climate, but it needs to be used with real care.

Project Evaluation

At the end of a project it is important, and often a requirement, to conduct an evaluation. Typically a set of project evaluation questions guides this process. The following, generated by Haynes (1989), is a typical set of questions:

1. How close to scheduled completion was the project actually completed?
2. What did we learn about scheduling that will help us on our next project?
3. How close to budget was final project cost?
4. What did we learn about budgeting that will help us on our next project?

5. Upon completion, did the project output meet client specifications without additional work?
6. If additional work was required, please describe.
7. What did we learn about writing specifications that will help us on our next project?
8. What did we learn about staffing that will help us on our next project?
9. What did we learn about monitoring performance that will help us on our next project?
10. What did we learn about taking corrective action that will help us on our next project?
11. What technological advances were made?
12. What tools and techniques were developed that will be useful on our next project?
13. What recommendations do we have for future research and development?
14. What lessons did we learn from our dealings with service organizations and outside vendors?
15. If we had the opportunity to do the project over, what would we do differently?

Continual Evaluation

Evaluation doesn't have to occur only at the end of the project; it is often initiated when a project falls behind schedule or goes over budget. My recommendation is that evaluation be an integral part of the project. You have probably been involved in group projects that got behind schedule or use more resources that were initially allocated.

There are lots of internal things you can do with your project team to address delays and resource excesses. Sometimes it's necessary to try to change external conditions to address delays and overruns. Here's a list of some things you can do:

INDIVIDUAL REFLECTION How have you dealt with projects that got behind schedule or required more resources than were initially allocated? What are some of the strategies you've used? Take a few minutes and reflect on dealing with delays and cost overruns.

1. *Recover later in the project.* If there are early delays or overruns, review the schedule and budget for recovery later. This is a common strategy in many projects. How often have you done extraordinary work at the last minute, especially the night before the project is due?

2. *Reduce project scope.* Consider eliminating nonessential elements or containing scope creep. Engineers often find better ways of doing things during projects and are sometimes perfectionists, so there is a tendency for the scope to creep.

3. *Renegotiate.* Discuss with the client the possibility of extending the deadline or increasing the budget. How often have you asked a professor for an extension? This is a common strategy, but sometimes there is not flexibility.

4. *Add additional resources.* Adding more resources—people, computers, and so on—to a project (activities along the critical path, as you learned in Chapter 6) can reduce the duration. The increased costs must be traded off with the benefits of the reduced schedule.

5. *Offer incentives or demand compliance.* Sometimes by offering incentives (provided you don't endanger people's lives or sacrifice performance specifications) you can get a project back on track. Other times you may have to demand that people do what they said they would do.

6. *Be creative.* If the delay is caused by resources that have not arrived, you may have to accept substitutions, accept partial delivery, or seek alternative sources.

Building Quality into Projects

Evaluation and continual improvement often become an ongoing part of projects and company culture. This aspect of company or organizational culture is commonly described as a quality initiative. Table 7.3 provides a set of insightful contrasts about old and new thinking about quality.

Engineers are often required to help develop a quality initiative in their organization. You may been involved in a quality initiative in your work or school. Although there has been some attention paid to quality in schools (see Langford and Cleary, 1995, for example), much of the emphasis on quality has been in the workplace. Business and industry have taken the lead, as indicated by Ford Motor Company's motto "Quality Is Job One."

Some quality basics include a systems perspective, emphasis on the customer, and understanding variation. It is essential to have knowledge of sources of variation, especially ways of measuring and documenting them, and strategies for reducing variability and maintaining consistent quality.

The current quality movement started with work by W. Edwards Deming in the 1940s and progressed through a series of initiatives—continuous quality improvement (CQI), total quality management (TQM), and presently, Six Sigma. Each successive phase seems to provide enhanced features for focusing on quality. These are the essential themes of Six Sigma (Pande, Neuman, and Cavanagh, 2000):

- A genuine focus on the customer
- Data- and fact-driven management
- Process focus, management and improvement, as an engine for growth and success
- Proactive management
- Boundaryless collaboration
- A drive for perfection, and yet a tolerance for failure

Another central feature of Six Sigma is the Six Sigma improvement model DMAIC—Define, Measure, Analyze, Improve, Control. Many students (especially those who work for 3M, Honeywell, Seagate, Medtronic, and so forth) in the Management of Technology master's program course I teach have in their

Table 7.3 Thinking About Quality

Old	New
Competition motivates people to do better work	Cooperation helps people do more effective work
For every winner there's a loser	Everyone can win
Please your boss	Please your customer
Scapegoating pinpoints problems	Improve the system
Focus improvements on individual processes	Focus on the purpose of the overall system, and how the processes can be improved to serve it better
Find the cause and fix the problem	First, acknowledge there is variation in all things and people. See if the problem falls in or outside the system
The job is complete if specifications have been met	Continual improvement is an unending journey
Inspection and measurement ensure quality	A capable process, shared vision and aim, good leadership and training are major factors in creating quality
Risks and mistakes are bad	Risks are necessary and some mistakes inevitable when you practice continual improvement
You can complete your education	Everyone is a lifelong learner
Bosses command and control	Bosses help workers learn and make improvements
Bosses have to know everything	The team with a good leader knows and can do more
Short-term profits are best	Significant achievement in a complex world takes time
You don't have to be aware of your basic beliefs	You must be conscious of your beliefs and constantly examine and test them to see if they continue to be true
Do it now	Think first, then act

Source: Dobyns and Crawford-Mason, 1994.

bag a copy of Rath and Strong's *Six Sigma Pocket Guide* (2000), which is based on the DMAIC model.

Further readings on quality can be found in the references in Deming (1993, 1986), Bowles and Hammond (1991), Dobyns and Crawford-Mason (1994), Brassand (1989), Sashkin and Kiser (1993), and Walton (1986). You may also want to consult a basic textbook on quality, such as Summers (1997).

Reflection: Paying Attention

I am often reminded of the advice of the talking mynah birds in Aldous Huxley's *Island* (1962). These birds periodically called out "Attention, attention; Here and now." Huxley is emphasizing awareness, as indicated in the following dialogue:

> "Listen to him closely, listen discriminatingly. . . ." Will Farnaby listened. The mynah had gone back its first theme. "Attention," the articulate oboe was calling. "Attention." "Attention to what?" he asked, in the hope of eliciting a more

enlightening answer than the one he had received from Mary Sarojini. "To attention," said Dr. MacPhail. "Attention to attention?" "Of course." (p. 16)

We often get so wrapped up in achieving our goals and completing our tasks that we forget to pay attention to what's going on around us. My sense is that successful project managers have refined their skills for paying attention.

Questions

1. Where did you develop skills for monitoring the work on project teams? Have you had an opportunity to observe a project team? If so, where? What did you learn from the experience? Do you try to attend to what's happening within the group while it is working?
2. What can you do to improve the functioning of teams during "boring and useless" meetings? List things you can do and strategies for doing them. Try them out.
3. Check out some of the ethnographic research on work in organizations, such as Brown and Duguid (1991). This paper is also available on the Xerox Palo Alto Research Center website. How does research affect your view of work in organizations?
4. What are some strategies for building quality into projects?
5. How can project evaluation, which is often seen as a punitive process, become a more positive and constructive process? What things can project team members and managers do to make evaluation an ongoing part of project work?

References

Bennis, W., and P. W. Biederman. 1997. *Organizing genius: The secrets of creative collaboration.* Reading: Addison-Wesley.

Bowles, J., and J. Hammond. 1991. *Beyond quality: How 50 winning companies use continuous improvement.* New York: Putnam.

Brassand, Michael. 1989. *The memory jogger plus+: Featuring the seven management and planning tools.* Methuen, MA: GOAL/QPC.

Brown, John Seely, and Paul Duguid. 1991. Organizational learning and communities-of-practice: Toward a unified view of working, learning, and innovation. *Organizational Science* 2(1): 40–56.

Bucciarelli, Louis L. 1996. *Designing engineers.* Cambridge, MA: MIT Press.

Cuff, Dana. 1991. *Architecture: The story of practice.* Cambridge, MA: MIT Press.

Deming, W. E. 1986. *Out of crisis.* Cambridge, MA: MIT Center for Advanced Engineering Study.

Deming, W. E. 1993. *The new economics for industry, government, education.* Cambridge, MA: MIT Center for Advanced Engineering Study.

Dobyns, Lloyd, and Clare Crawford-Mason. 1994. *Thinking about quality: Progress, wisdom, and the Deming philosophy.* New York: Times Books.

Donnellon, Anne. 1996. *Team talk: The power of language in team dynamics.* Cambridge, MA: Harvard Business School Press.

Haynes, M. E. 1989. *Project management: From idea to implementation.* Los Altos, CA: Crisp.

Huxley, Aldous. 1962. *Island.* New York: Harper & Row.

Katzenbach, Jon R., and Douglas K. Smith. 1993. *The wisdom of teams: Creating the high-performance organization.* Cambridge, MA: Harvard Business School Press.

Langford, David P., and Barbara A. Cleary. 1995. *Orchestrating learning with quality.* Milwaukee: ASQC Quality Press.

Leeper, John. 2002. Personal communication. November 7.

Lewis, J. P. 1998. *Team-based project management.* New York: AMACOM.

Lippincott, S. 1994. *Meetings: Do's, don'ts, and donuts: The complete handbook to successful meetings.* Pittsburgh: Lighthouse Point Press.

McGourty, Jack, and Kenneth P. De Meuse. 2001. *The team developer: An assessment and skill building program.* New York: Wiley.

Minneman, Scott. 1991. *The social construction of a technical reality: Empirical studies of group engineering design practice.* Xerox Corporation Palo Alto Research Center Report SSL-91-22. Palo Alto, CA: Xerox Corporation.

Pande, Peter S., Robert P. Neuman, and Roland D. Cavanagh. 2000. *The six sigma way: How GE, Motorola and other top companies are honing their performance.* New York: McGraw-Hill.

Peters, Thomas J. 1989. *A passion for excellence.* (With Nancy Austin). New York: Knopf.

Peters, Thomas J. 1999. The WOW project: In the new economy, all work is project work. *Fast Company,* May, 116–128.

Rath, and Strong. 2000. *Six sigma pocket guide.* Lexington, MA: Rath and Strong Management Consultants.

Sashkin, Marshall, and Kenneth J. Kiser. 1993. *Putting total quality management to work.* San Francisco: Berrett-Koehler.

Scholtes, Peter. 1988. *The team handbook: How to use teams to improve quality.* Madison, WI: Joiner Associates.

Scholtes, P. R., B. L. Joiner, and B. J. Streibel. 1996. *The team handbook.* Madison: Joiner Associates.

Summers, Donna C. S. 1997. *Quality.* Upper Saddle River, NJ: Prentice Hall.

Uhlfedler, H. 1997. Ten critical traits of group dynamics. *Quality Progress* 30(4): 69–72.

Walton, M. 1986. *The Deming management method.* New York: Putnam.

Welch, John F., Jr., and John A. Byrne. 2001. *Jack: Straight from the gut.* New York: Warner.

Wenger, E., and W. Snyder. 2000. *Communities of practice: The next organizational frontier. Harvard Business Review.*

Xerox Corporation. 1986. *Leadership through quality: Interactive skills workbook.* Stamford, CT: Author.

CHAPTER 8

Project Management Documentation and Communications

"A horse, a horse! My kingdom for a horse," cried Richard III in Shakespeare's play of the same name. Although most of us no longer need horses to travel, we need good documentation for successful projects, lest we find ourselves crying, "Good documentation, good documentation! My career for good documentation."

This chapter is organized into two sections—Project Documentation and Project Communications. Project documentation is stressed because it is often neglected and because there are many other fine resources on project communication (i.e., reports and presentations), such as *A Beginner's Guide to Technical Communication,* by Anne Eisenberg (1998).

Project Documentation

Why are project documentation and record keeping important? To answer this question you only have to think about a time when you had to pick up the pieces from another team member and pull them together into a report or presentation. Was everything easy to follow and understand, or did you have to fill in or reconstruct much of the work?

Problem solving, especially if it involves developing a computer program, is particularly susceptible to gaps in documentation. Think about how well you've documented projects you've been involved in. Did you insert lots of comments in your programs or spreadsheets to let the next user know what you'd done and why?

Many of the problem-solving and program-writing assignments students hand in lack sufficient documentation; thus, the faculty member cannot assess whether the procedures are correct (even if the answer isn't). The lack of documentation makes it very difficult for faculty to grade such reports. In engineering practice, poor documentation can make it difficult for a client to understand the work, and might therefore jeopardize future contracts.

Leifer's (1997) research at the Stanford University Center for Design Research indicates that "all design is redesign." The more information and insight you provide those who follow you, the better job they will be able to do. Of course, this works for you, too, when you follow up on someone else's work. Project documentation is important to everyone involved in the project.

REFLECTION Take a few minutes to think about the types of project records that should be kept. Make a list. Next, think about the characteristics of good records. List several attributes of good records. Compare your list with team members' lists.

What did you come up with? How easy or hard was it for you to think about types of documents and the characteristics of good documentation?

Many engineers prefer to focus on solving the problem, developing the product, or just getting the job done rather than on documentation, which they see as a necessary (and often unpleasant) burden.

For comparison, here are lists that were developed by the participants in the Minnesota Department of Transportation Project Management Academy:

Types of Records

Formal	**Informal**
Specifications	Survey notes
Drawings	Inspection reports
Schedules	Photographs/videotapes
Budgets	Notes—personal and meeting
Contracts	Incident reports
	Telephone/e-mail/FAX memos and notes
	Commitment logs
	Change orders
	On-site log—date/weather/time/personnel/ equipment

Characteristics of Good Records

Easily accessible	Right media
Thorough—date and time, client	Consistent format
Organized and legible	Secure
Comprehensive—table of contents	Cost effective
	Flexible

Nearly two hundred forms were recently compiled by the Project Management Institute (PMI) in a book titled *Project Management Forms* (Pennypacker, Fisher, Hensley, and Parker, 1997). PMI members shared their forms, checklists, reports, charts, and other documents to help you get started or to improve your current documentation. More information is available on the PMI website (www.pmi.org).

Notebooks and Journals

Notebooks and journals are terrific ways to document work for your own personal use, and there are many examples of their significance in patent applications and even Nobel Prize considerations. Think, for example, how the notebooks of Charles Kettering (inventor of the electric cash register and automobile ignition systems) and Shockley, Brattain, and Bardeen (inventors of the transistor) helped establish intellectual property rights. Also consider that Bill Gates paid $30 million for Leonardo da Vinci's *Codex Leicester,* one of da Vinci's surviving journals.

Students in ERG 291, a freshman design course at Michigan State University, are required to keep a laboratory notebook/academic journal. The box "Academic Journal" describes the documentation requirements.

Academic Journal

What Is a Journal? A journal is a place to practice writing and thinking. It differs from a diary in that it should not be merely a personal recording of the day's events. It differs from your class notebook in that it should not be merely an objective recording of academic data. Think of your journal rather as a personal record of your educational experience in this class. For example, you may want to use your journal while working on a design project to record reflections on the class.

What to Write. Use your journal to record personal reactions to class, topics, students, teachers. Make notes to yourself about ideas, theories, concepts, problems, etc. Record your thoughts, ideas, and readings, and to argue with the instructor, express confusion, and explore possible approaches to problems in the course. Be sure to include (1) critical incidents that helped you learn or gain insight, and (2) distractions that interfered with your learning.

When to Write. Try to write in your journal at least three or four times a week . . . aside from classroom entries. It is important to develop the *habit* of using your journal even when you are not in an academic environment. Good ideas, questions, etc., don't always wait for convenient times for you to record them. [*A man would do well to carry a pencil in his pocket and write down the thoughts of the moment. Those that come unsought for are the most valuable and should be secured because they seldom return. Francis Bacon (1561–1626)*].

How to Write. You should write in a style that you feel most comfortable with. The point is to think on paper without worrying about the mechanics of writing. The quantity you write is as important as the quality. The language that expresses your personal voice should be used. Namely, language that comes natural to you.

Suggestions:

1. Choose a notebook you are comfortable with. A 8½" × 11" hardback bound book with numbered pages would be a good choice.
2. Date each entry including time of day.
3. Don't hesitate to write long entries and develop your thoughts as fully as possible.
4. Include sketches and drawings.
5. Use a pen.
6. Use a new page for each new entry.
7. Include both "academic" and "personal" entries; mixed or separate as you desire.

Interaction—Professor. I'll ask to see your journal at least twice during the term: I'll read selected entries and, upon occasion, argue with you or comment on your comments. Mark any entry that you consider private and *don't want me to read* and I'll gladly honor your request. A *good* journal will contain numerous long entries and reflective comments. It should be used regularly.

Interaction—Correspondent. Choose a fellow student in your close collection of friends to read and respond to your journal entries.

Conclusions. Make a table of contents of significant entries. At the end of the semester write a two-page summary. In addition, submit an evaluation of whether the journal enhanced or detracted from your learning experience. Was it worth the effort?

Source: Adapted from Fulwiler, T. 1987. *Teaching with writing.* Portsmouth, NH: Boynton/Cook Publishers. Revised by B. S. Thompson in consultation with K. A. Smith and R. C. Rosenberg, 1998.

Chapter 7, "Information in Design," of *Understanding Engineering Design* (Birmingham, Cleland, Driver, and Maffin, 1997) begins with the following sharply focused statement on the importance of information: "The raw material of the design process is information, and therefore the designer's principal skill is one of information handling" (p. 108). The authors stress five categories of action that operate on information:

1. Collection
2. Transformation
3. Evaluation
4. Communication
5. Storage

Furthermore they stress that these categories are used at all stages of the design process.

One of the challenges of personal notebooks and journals is that they are not easily shareable. Lack of easy access to others' work and thinking makes for considerable problems in team-based project work. Larry Leifer and his colleagues at Stanford have devised an electronic notebook that is accessible on the World Wide Web. The Personal Electronic Notebook with Sharing (PENS) (Hong, Toye, and Leifer, 1995) supports and implements Web-mediated selective sharing of one's working notes. Electronic mail is another common way to share thinking about projects. Numerous software products, such as Lotus Notes, provide a means for sharing information electronically. Many electronic calendars and personal data assistants (PDAs) have features for jointly scheduling meetings by viewing others' calendars. PDAs and electronic calendars also provide excellent means for keeping records.

The image in Figure 8.1 is from the desktop interface of my electronic calendar, a Palm Pilot. I'm currently using a Palm IIIc to organize and keep track of appointments, addresses, and my todo list. In addition to having several calendar options, it also has an address book, to-do list, memo pad, and expense section. PDAs such as the Palm provide for portability, backing up, and electronic sharing. In addition, all of the sections can be quickly searched, which makes for easy information retrieval. The PDA will likely become an essential tool for project managers and are currently being used on construction projects (Roe, 2001). Computer-based tools are discussed further in Chapter 9.

Project Communications

Presenting our ideas to others in both written and oral form is essential to successful projects. You may, however, feel a combination of excitement and anxiety about report writing and public presentation.

Some of the best advice I ever got on presentations is to know your audience, know your objective, and be simple, concise, and direct. This brief section follows that advice.

Communications need to be tailored to the recipients, whether they be your colleagues on the team, your supervisor, or your client. One way to tailor communications is to learn more about the recipients through surveys, interviews, or informal conversations.

Figure 8.1 Palm PDA Desktop View.

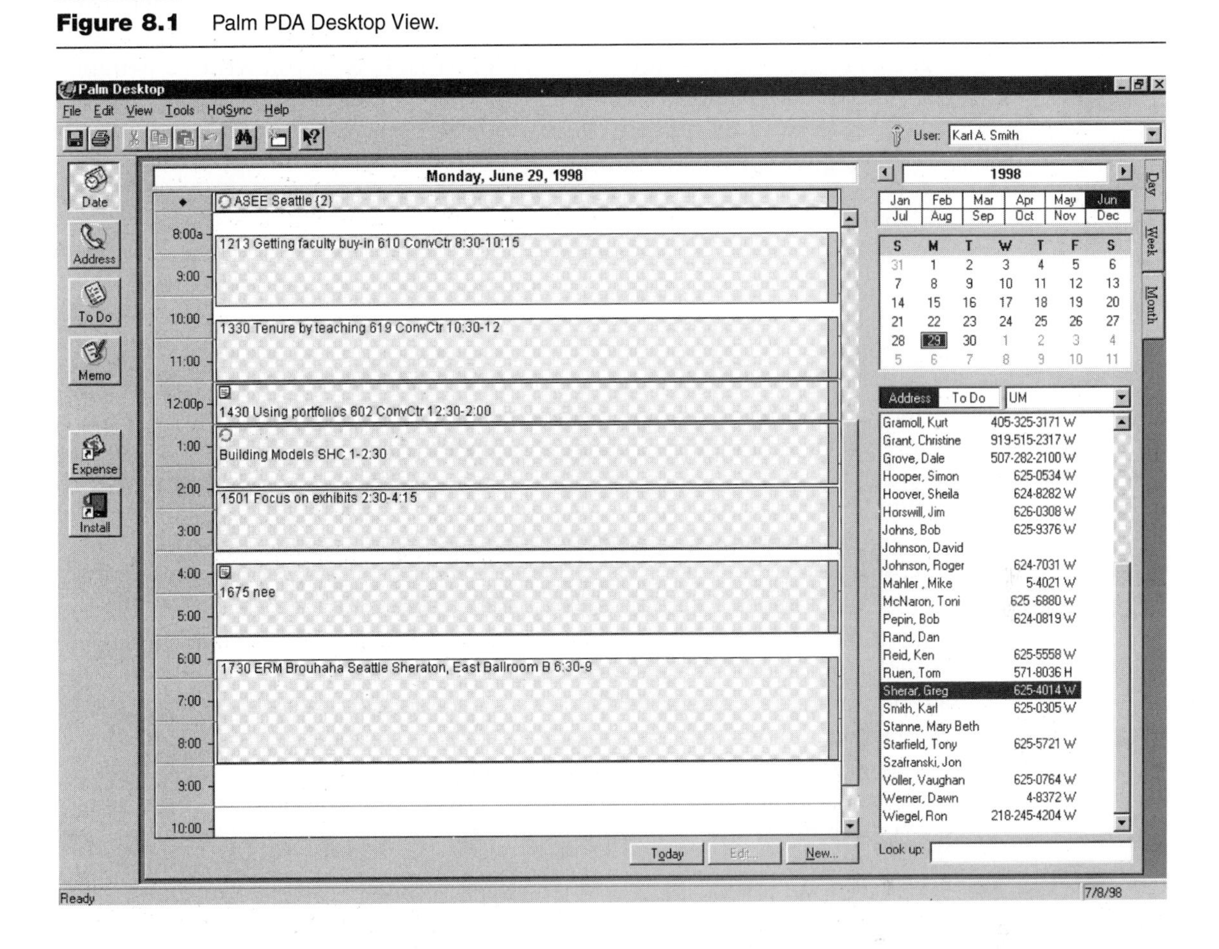

REFLECTION Take a minute to reflect on some possible objectives in a presentation or report. Try to think beyond "Because it's an assignment in this class." As much as possible, tie your list to actual experiences that you've had. If you have a chance, discuss these with other team members.

One of the best references on simple, concise, and direct writing is *The Elements of Style* by Strunk and White (1979). One of my other favorites is Williams's (1997) *Style: Ten Lessons in Clarity and Grace.*

Getting Started, Keeping Going

A practice commonly called freewriting may help you get over the barrier that many of us face when writing or preparing for an oral presentation. A helpful implementation of freewriting is Natalie Goldberg's advice for "writing practice" in her book *Writing Down the Bones: Freeing the Writer Within* (1986).

1. *Keep your hand moving.* (Don't pause to reread the line you have just written. That's stalling and trying to get control of what you're saying.)
2. *Don't cross out.* (That is editing as you write. Even if you write something you didn't mean to write, leave it.)
3. *Don't worry about spelling, punctuation, grammar.* (Don't even care about staying within the margins and lines on the page.)
4. *Lose control.*
5. *Don't think. Don't get logical.*
6. *Go for the jugular.* (If something comes up in your writing that is scary or naked, dive right into it. It probably has lots of energy.)

Revising and Refining

After you complete a draft of your paper or presentation, share it with others. Many papers and reports must undergo a formal review process, but I recommend that you offer your work for informal review with a colleague or friend. Although it's hard for many of us to share our work with others before we feel it is ready, we can often save a lot of time and get much better products by asking others for comments.

At some stage in the process, of course, you must freeze the design and submit the report or give the presentation. Be sure to solicit feedback (in a constructive mode), reflect on it, and consider changes for the next time you write a report or give a presentation.

Leaving an excellent paper and electronic record is likely to increase in importance in projects and project management. Now is the time to develop skills and strategies for effective documentation. Similarly, communicating effectively both orally and in writing will continue to be extremely important for your project success.

Reflection: Documentation

During the 20 or so years I worked in an engineering research lab, I was required to keep a laboratory notebook. Although I had kept a notebook as a graduate student, I was struck by the seriousness and formality of the lab's research notebooks. Laboratory notebooks were numbered and signed out. Each page was numbered and had a duplicate page (for a carbon copy). Duplicate pages were collected periodically and filed in a safe place. My mentor in this process, Dr. Iwao Iwasaki, would ask me to notarize his signature every week or so. This was to certify the date and signature for intellectual property rights and for eventually filing patent disclosures and applying for patents. Several of the ideas he, and later I, signed off on were developed as patent disclosures. Dr. Iwasaki's ideas and research earned him membership in the National Academy of Engineering, the most prestigious engineering organization in the United States. It was always difficult for me to take the time to write everything down, but it made a huge difference when we were working on reports and publications and were trying to figure out exactly what we did and what we found. The notebooks were much better than my memory.

In terms of report preparation, I recommend that you get a copy of a good technical communications book, such as the ones I described before (I've used Houp and Pearsall, 1980, throughout my career) and use it every time you work on a report writing assignment.

Questions

1. What is your experience documenting group projects? Is it a routine activity? If so, describe examples of "excellent documentation." If not, consider how can you build the development of good documentation into the ongoing process of project work.
2. What are some of the types of records that must be maintained for projects? What are the characteristics of good records?
3. What is your experience with electronic documentation? What are the advantages and disadvantages of electronic records (compared with paper records)?
4. Describe the characteristics of good presentations. Are good presentations the norm in your experience? Why or why not?
5. Describe your experiences keeping an academic journal. What are some of the heuristics that helped make it effective for you?
6. Learning to become a better writer and presenter requires effort and practice. How are you planning to improve your writing and presenting skills? What are some of your favorite resources?

References

Birmingham, R., G. Cleland, R. Driver, and D. Maffin. 1997. *Understanding engineering design: Context, theory and practice.* London: Prentice Hall Europe.

Eisenberg, A. 1998. *A beginner's guide to technical communication.* New York: McGraw-Hill.

Goldberg, N. 1986. *Writing down the bones: Freeing the writer within.* Boston: Shambhala Press.

Hong, J., G. Toye, and L. Leifer. 1995. Personal Electronic Notebook with Sharing. In *Proceedings of the IEEE Fourth Workshop on Enabling Technologies: Infrastructure for Collaborative Enterprises (WET ICE).* Berkeley Springs, WV.

Houp, Kenneth W., and Thomas E. Pearsall. 1980. *Reporting technical information.* New York: Macmillian.

Leifer, L. 1997. *A collaborative experience in global product-based-learning.* Paper presented at the National Technological University Faculty Forum. November 18, Boulder, Colorado.

Pennypacker, James S., Lisa M. Fisher, Bobby Hensley, and Mark Parker. 1997. *Project management forms.* New Square, PA: Project Management Institute.

Roe, Arthur G. 2001. Handhelds hold up well for a variety of site uses. *ENR: Engineering News-Record,* November 11, 31–32.

Strunk, W., Jr., and E. B. White. 1979. *The elements of style,* 3rd ed. New York: Macmillan.

Swain, Philip W. 1945. Giving power to words. *American Journal of Physics* 13(5): 320.

Thompson, B. S. 1997. *Creative engineering design.* Okemos, MI: Okemos Press.

Williams, J. M. 1997. *Style: Ten lessons in clarity and grace.* New York: Longman.

CHAPTER 9

Project Management Software

The more time we spend on planning a project, the less total time is required for it. Don't let today's busywork schedule crowd planning time out of your schedule.

EDWARD BLISS
Getting Things Done

A wide variety of software tools is available to help the project manager and project team members accomplish their goals. These include personal data assistants (PDAs), which include electronic calendars, address books, to-do lists, memo pads, and sometimes expense reports; project management programs that do scheduling, resource leveling, and tracking; and Web-based project enterprise systems that manage both of the previous functions as well as drawings and specifications, budgets, contracts, and much more.

Personal Data Assistants

Personal data assistants (PDAs) help project managers keep track of appointments, critical deadlines, to-do lists, notes, and expenses. Many provide for access to calendars over a network or over the Internet, a feature that makes it much easier to schedule meetings. Most provide for synchronizing of data between the PDA and a computer, which makes it possible to easily take the information into the field. See Figure 8.1 for a sample screen from the Palm Pilot desktop. Paper calendars and planners are inexpensive, but they cannot be backed up easily (except by photocopying), nor can the information be shared with others very easily (which has its advantages).

> INDIVIDUAL REFLECTION What type of calendar (or planner) are you currently using? Is it a small paper datebook? a leather-bound three-ring binder? Or are you using a PDA or computer-based planner? What are the principal uses that you make of your planner?

Personal data assistants are mainly used to manage time, priorities, and contacts. They help project managers attend to the details that are crucial for successful teamwork and project management. As PDAs become more powerful, and with the advent of wireless communication, they are being used for inspection and project check-off, which results in much quicker turnaround (Bryant and Pitre, 2003a; Roe, 2001).

Project Management Software

Comprehensive project management software such as Microsoft Project is used on complex projects to accomplish goals and complete projects on time, within budget, at a level of quality that meets the client's expectations. The basic functions of the critical path analysis aspect of these programs in summarized in Chapter 6, where the CritPath program is featured.

The two most common views used by commercial project management software are the Gantt chart and precedence network chart, sometimes referred to as a PERT chart. Figures 9.1 and 9.2 show examples of the Gantt chart and PERT chart views, respectively, from Microsoft Project.

Examples of the Gantt chart and PERT chart views from Primavera are shown in Figures 9.3 and 9.4, respectively. These Primavera views show the activity detail for a highlighted activity; this allows the project manager to quickly get lots of detailed information on any activity, which makes it easy to track, manage, and update information.

Microsoft Project and Primavera are the two most widely used project management software packages. In a survey regarding project management tools

Figure 9.1 Microsoft Project—Gantt Chart.

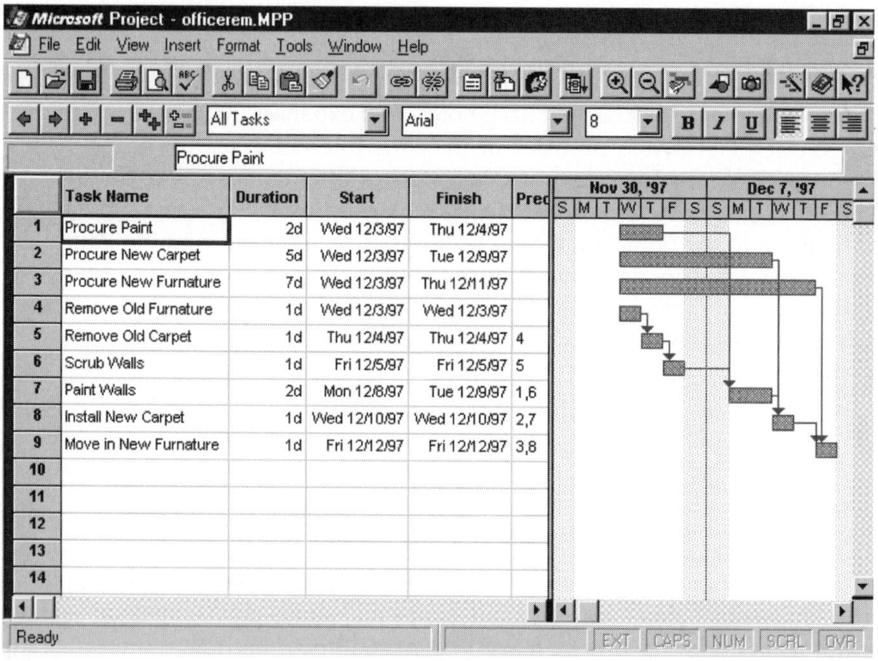

Microsoft Project - officerem.MPP

File Edit View Insert Format Tools Window Help

All Tasks | Arial | 8

Procure Paint

	Task Name	Duration	Start	Finish	Pred
1	Procure Paint	2d	Wed 12/3/97	Thu 12/4/97	
2	Procure New Carpet	5d	Wed 12/3/97	Tue 12/9/97	
3	Procure New Furnature	7d	Wed 12/3/97	Thu 12/11/97	
4	Remove Old Furnature	1d	Wed 12/3/97	Wed 12/3/97	
5	Remove Old Carpet	1d	Thu 12/4/97	Thu 12/4/97	4
6	Scrub Walls	1d	Fri 12/5/97	Fri 12/5/97	5
7	Paint Walls	2d	Mon 12/8/97	Tue 12/9/97	1,6
8	Install New Carpet	1d	Wed 12/10/97	Wed 12/10/97	2,7
9	Move in New Furnature	1d	Fri 12/12/97	Fri 12/12/97	3,8
10					
11					
12					
13					
14					

Nov 30, '97 | Dec 7, '97

S M T W T F S S M T W T F S

Ready | EXT | CAPS | NUM | SCRL | OVR

Figure 9.2 Microsoft Project—PERT Chart.

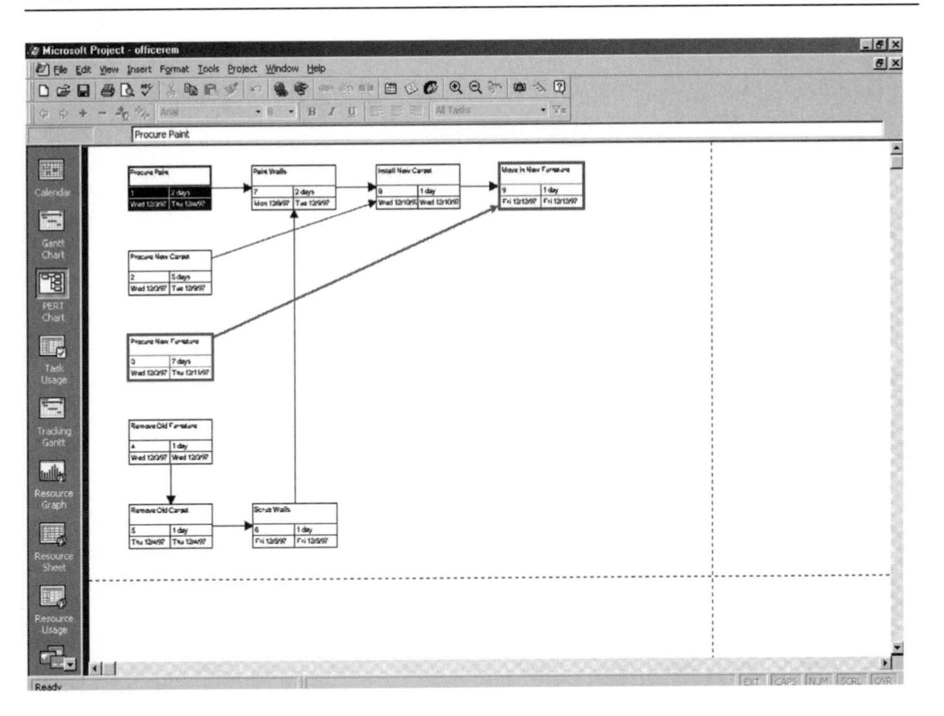

Figure 9.3 Primavera—Gantt Chart.

Primavera Project Planner - [BUS1]

File Edit View Insert Format Tools Data Window Help

Roof beam — 01DEC97 Mon

Activity ID	Activity Description	Orig Dur	Rem Dur	Early Start	Early Finish
5	Excavation	2	2	02DEC97	03DEC97
4	Roof beam	3	3	02DEC97	04DEC97
6	Curb and gutter	2	2	02DEC97	03DEC97
1	Shelter slab	2	2	04DEC97	05DEC97
7	Shelter Seat	1	1	05DEC97	05DEC97
2	Shelter walls	1	1	06DEC97	06DEC97
8	Paint	1	1	06DEC97	06DEC97
3	Shelter roof	2	2	07DEC97	08DEC97

1997 DEC 1 2 3 4 5 6 7 8 9 10 11

Excavation / Roof beam / Curb and gutter / Shelter slab / Shelter Seat / Shelter walls / Paint / Shelter roof

Budget | Codes | Constr | Cost | Custom | Dates | Log | Pred | Res | Succ | WBS

ID 4 Roof beam — Previous Next Help

OD 3 Pct 0.0 Cal 1 — ES 02DEC97 — EF 04DEC97 — TF: 2

RD 3 Type Task — LS 04DEC97 — LF 06DEC97 — FF: 0

Resp Area Mile Item Locn Step WBS

Classic Schedule Layout — All Activities

Figure 9.4 Primavera—PERT Chart.

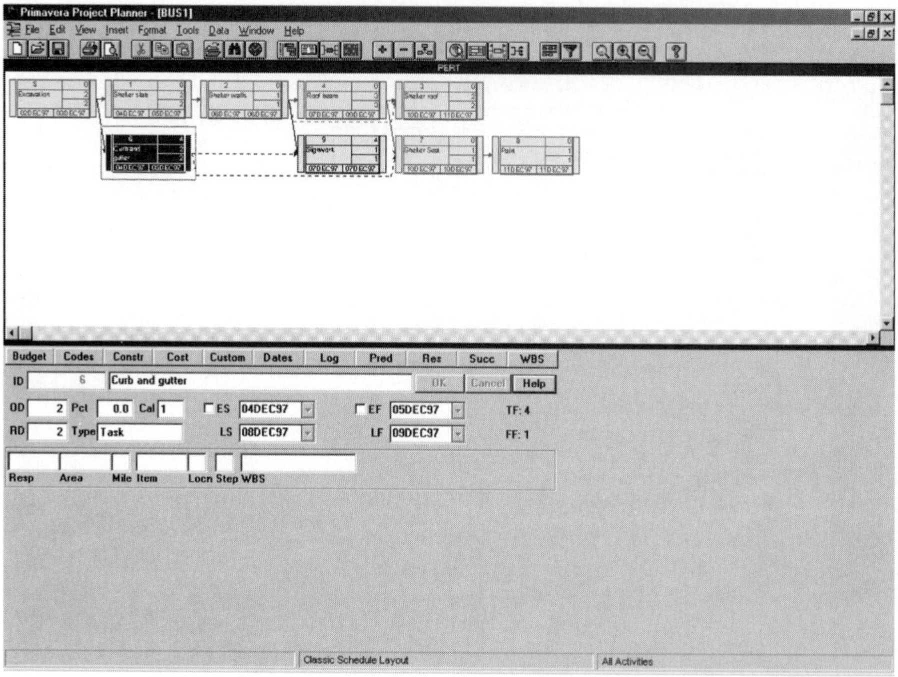

by Fox and Spence (1998), 48 percent of the respondents reported using Microsoft Project and 14 percent reported using Primavera.

Pollack-Jackson and Liberatore (1998) reported similar figures—nearly 50 percent for Microsoft Project and 21 percent for Primavera—and provided extensive information on how these packages are used. The median size of projects reported by the respondents was a little over 150 activities, and the median number of resources was 16. A high percentage of respondents reported that they regularly update the information, and about 62 percent of the respondents said they use resource scheduling/leveling.

On a more personal (and perhaps mundane) level, I now include a Gantt chart summarizing the assignments in my project management and economics syllabus (see Figure 9.5). This visual approach helps many students internalize the schedule so they know when they need to be working on the various assignments, and it especially focuses them on due dates.

The Project Management Institute recently launched a major project management software survey, available both in print and on CD-ROM (Cabanis, 1999). The survey

- Compares and contrasts the capabilities of a wide variety of project management tools.
- Provides a forum for users and vendors to meet and match requirements and possibilities.

Figure 9.5 Project Management Course Assignments—Gantt Chart.

ID	Task Name	Duration	Start	Finish
1	Classes Begin	0 days	Tue 9/3/02	Tue 9/3/02
2	Homework 1	6 days	Thu 9/5/02	Thu 9/12/02
4	Brief Review 1	11 days	Thu 9/12/02	Thu 9/26/02
3	Homework 2	11 days	Tue 9/24/02	Tue 10/8/02
5	Group Project 1	16 days	Tue 9/24/02	Tue 10/15/02
8	Brief Review 2	13 days	Tue 10/1/02	Thu 10/17/02
16	Exam 1	0 days	Thu 10/10/02	Thu 10/10/02
6	Homework 3	6 days	Tue 10/15/02	Tue 10/22/02
11	Project Proposal	19 days	Thu 10/17/02	Tue 11/12/02
9	Group Project 2	13 days	Tue 10/29/02	Thu 11/14/02
7	Homework 4	6 days	Thu 11/14/02	Thu 11/21/02
13	Major Paper	21 days	Thu 11/14/02	Thu 12/12/02
17	Exam 2	0 days	Tue 11/19/02	Tue 11/19/02
10	Delta Design	4 days	Thu 11/21/02	Tue 11/26/02
12	Thanksgiving Break	2 days	Thu 11/28/02	Fri 11/29/02
15	Homework 5	6 days	Tue 12/3/02	Tue 12/10/02
14	Last Day of Instruction	0 days	Fri 12/13/02	Fri 12/13/02
18	Exam 3	0 days	Thu 12/19/02	Thu 12/19/02

- Prompts vendors to become more responsive to customer needs.
- Prompts users to create a methodology for software tool selection within their own companies.
- Categorizes software tools into six areas of functionality aligned with the knowledge areas of PMI's *Guide to the Project Management Body of Knowledge (PMBOK Guide)*: scheduling, cost management, risk management, human resources management, communications management, and process management (Project Management Institute, 2000).

Like all software tools, it is important that project management software serve and not enslave the project manager. Also, if you invest the time and money in commercial project management software, you should use it to organize and manage your projects and not simply as a reporting tool. Lientz and Rea (1995) offer the following suggestions for using project management software:

1. Set up the basic schedule information: name of project file, name of project, project manager; input milestones, tasks and their estimated duration, interdependencies between tasks; input resources for each task.
2. Periodically update the schedule by indicating tasks completed, delayed, and so forth, as well as changes in resources.
3. On an as-needed basis, perform what-if analysis using the software and data.

Uses of project management software include *reporting* (use the schedule to produce graphs and tables for meetings); *tracking* (log project work and effort in terms of completed tasks); *analysis* (perform analysis by moving tasks around, changing task interdependencies, changing resources and assignments, and then seeing the impact on the schedule); *costing and accounting* (assign costs to resources); and *timekeeping* (enter the time and tasks worked on by each member of the project team).

Unofficial reports indicate that more than 1 million copies of Microsoft Project have been sold. That's a lot of people scheduling projects. Advertisements for civil engineering positions often require that applicants be familiar with project management software, especially Primavera and Meridian Project Systems. Lots of books, short courses, and multimedia training programs are available to help you learn to use these tools. Some of the books I've found useful are included in the references (see Day, 1995; Lowery, 1994; Marchman, 1998). This is a rapidly changing area, so I suggest that you stay tuned to resources such as the Project Management Institute, especially via their website (www.pmi.org), to keep current.

> REFLECTION What's your experience with project management software packages, such as Microsoft Project? Has this software been bundled with any of your textbooks? Have you used these programs to schedule project work associated with work or school? Talk about this with your friends, and work to expand your repertoire of software tools.

Project Management and the World Wide Web

Project management has developed a Web presence. I'm currently working on three projects (two writing projects and a research project) where the repository for all the documents is eProject.com. The eProject home page and the My Proj-

Figure 9.6 eProject Home Page and My eProjects.

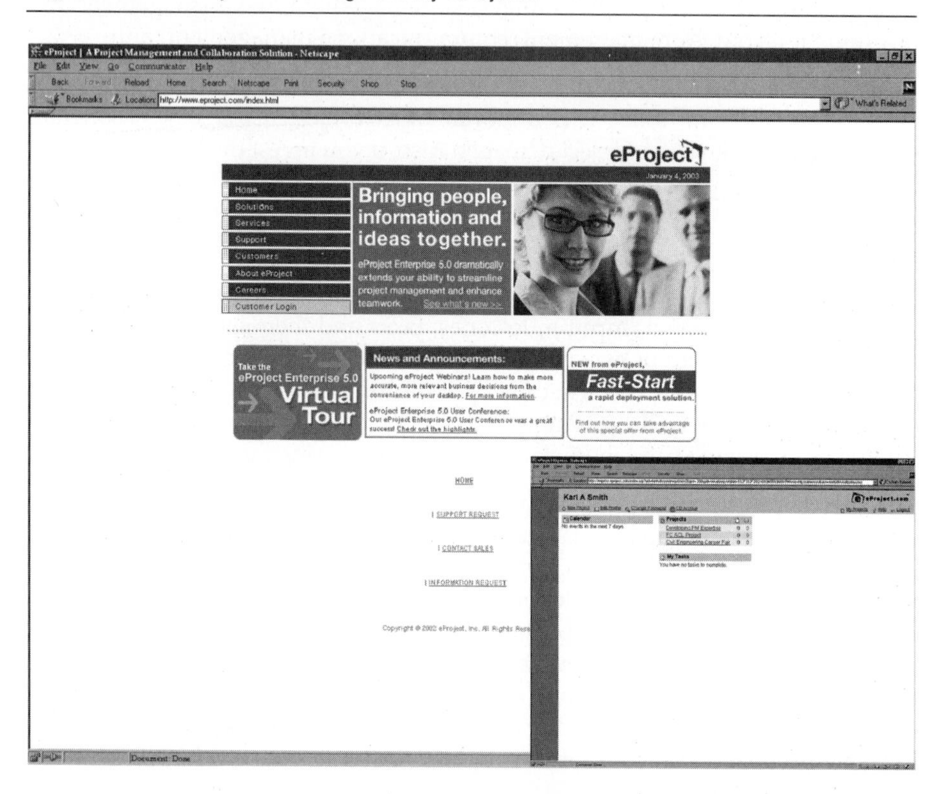

ects view on eProject are shown in Figure 9.6. The teams for these three projects are geographically distributed, and the eProject electronic repository and collaboratory provides us with a place to work together. Another kind of project for which this works well is grant writing. Members of the grant-writing team can be on different continents but share documents via an electronic repository. They might meet face-to-face periodically, but most or all of the work can be conducted in an electronically mediated environment.

As mentioned above, the Project Management Institute has a very thorough website and provides access to their *Guide to the Project Management Body of Knowledge* (*PMBOK Guide*). Like most other websites, project management websites update their URLs so frequently that if I had listed many here, most would already have been out of date by the time this book arrived at the bookstores. Use your favorite World Wide Web search engine (I'm partial to Google.com) and do a search on "project management." You'll probably be amazed at the number of useful and relevant sites that come up.

Bryant and Pitre (2003b) describe three broad categories of project management websites:

1. *Project collaboration networks (PCNs).* An environment for information sharing, such as eProject described above. An example in the construction industry is www.constructware.com.
2. *Project information portals (PIPs).* These contain essential project information on codes and specifications, permits, product information, cost data, and so forth. An example is www.aecdirect.com.
3. *Project procurement exchanges (PPEs).* These sites deal with the exchange of material, and are essential e-commerce. An example from the construction industry is www.buildpoint.com.

Web-enabled project management is gaining momentum and will probably change some businesses, just as it is currently changing the construction industry (Roe and Phair, 1999; Doherty, 1999). Web-enabled project management couples aspects involving communication (e-mail, fax, voice and multimedia, Intranet, Extranet, and others) and project management (scheduling, document and file management, project administration, job photos, job cost reports, and project status reports). Doherty (1999) cites several reasons for using a project Extranet:

- Fewer communication errors between project team members.
- Up-to-the-minute intelligence on all the decisions and collective information related to a project.
- Less expense for messengers, couriers, copying, and blueprints.
- Customized sites for each project and customized access for each user.
- Security.

Project management is about planning, scheduling, monitoring, and controlling projects, so there are enormous benefits to having a central project file located at a website rather than in a project notebook (or in the project manager's head). The challenge involves moving from our comfort zones of familiar practice and learning new tools and approaches.

Questions

1. Describe the advantages and disadvantages of different calendar/planner formats—pocket planner, three-ring binder, pocket electronic organizer, and computer-based personal data assistant.
2. What are the major types of project management software? What are their common uses?
3. What are the advantages and disadvantages of the Gantt chart view and the precedence network (PERT) views available in commercial project management software?
4. How could you apply Lientz and Rea's suggestions for using project management software to a project you're currently involved in?
5. Check out project management websites on the World Wide Web. Keep a record of your findings in a journal.

References

Bryant, John A., and Jyoti Pitre. 2003a. *Emerging technologies in the construction industry.* http://archnt2.tamu.edu/contech/finalJ/home.html (Accessed 1/4/03).

———. 2003b. *Web based project management.* http://archnt2.tamu.edu/contech/finalJ/web_based.html (Accessed 1/4/03).

Cabanis, Jeannette. 1999. *Project management software survey.* New Square, PA: Project Management Institute.

Day, Peggy. 1995. *Microsoft Project 4.0: Setting project management standards.* New York: Van Nostrand Reinhold.

Doherty, Paul. 1999. Site seeing. *Civil Engineering* 69(1): 38–41.

Feigenbaum, Leslie. 1998. *Construction scheduling with Primavera project planner.* Upper Saddle River, NJ: Prentice Hall.

Fox, Terry L., and J. Wayne Spence. 1998. Tools of the trade: A survey of project management tools. *Project Management* 29(3): 20–27.

Lientz, Bennet, and Kathryn Rea. 1995. *Project management for the 21st century.* San Diego: Academic Press.

Lowery, Gwen. 1994. *Managing projects with Microsoft Project 4.0.* New York: Van Nostrand Reinhold.

Marchman, David A. 1998. *Construction scheduling with Primavera project planner.* Albany, NY: Delmar.

Pollock-Jackson, Bruce, and Matthew J. Liberatore. 1998. Project management software usage patterns and suggested research directions for future development. *Project Management* 29(2): 19–28.

Project Management Institute. 2000. *A guide to the project management body of knowledge.* Project Management Institute.

Roe, Arthur G. 2001. Handhelds hold up well for variety of site uses. *ENR: Engineering News-Record,* November 5, pp. 31–32.

Roe, Arthur G., and Matthew Phair. 1999. Connection crescendo. *ENR: Engineering News-Record,* May 17, pp. 22–26.

CHAPTER 10

Where to Go from Here

Projects and teams are going to be with you for the rest of your life, no matter what profession you eventually work in. They are already prevalent in engineering, medicine, law, and most areas of business and industry. Even if you become a college professor, you will probably be involved in projects and teams, especially on research projects and with your graduate students. Now you have made a start at learning how to effectively participate in projects and teamwork. I hope you've also developed skills for managing and leading a team. There are lots of additional resources available, and I hope you will continue to read about teamwork and project management. More importantly, I hope you will talk with colleagues (fellow students and faculty) about teamwork and project management. If you aspire to become a project manager, I encourage you to check out the Project Management Institute. They have a special student membership rate, and your membership in this organization will help connect you with project management professionals. Most professional organizations, such as ASCE, ASME, and IEEE, have a division that emphasizes engineering management. Check these out as you become a student member of the professional organization in your discipline.

Periodically reflecting on your experiences, writing down your reflections (as I have asked you to do throughout this book), processing them alone and with others, and reading and studying further will help ensure that your team and project experiences are more constructive. A sustained effort will ensure that you continue to learn and grow.

If you are in a team or project situation that is not working well, rather than just endure it and hope it will pass quickly, try some of the ideas in this book for improving the team or project. Suggest that the team members discuss how effectively they are working together. For example, suggest a quick individually written plus/delta processing exercise to survey the team. Successful project work and teamwork does not happen magically; it takes continual attention not only to the task but also to how well the group is working together. And this is *work.* The satisfaction and sense of accomplishment that comes from effective team and project work is worth the effort. So many things can't be accomplished any other way. The more you learn during your college years, the easier it will be for you after graduation. Paying attention to these skills now will save you from what previous generations of engineering graduates have had to endure—learning project management and teamwork skills on the job in addition to all the other complex things they had to learn.

As you work with this book and the ideas and strategies for effective teamwork and project management, think about what else you need to know. Develop a learning and teaching plan for yourself and your team and project members. A few resources are listed in the references; these barely scratch the surface of all the resources that are available. Check out a few of them. Add your own favorites to the list. Share your list with colleagues. Several students I've worked with have found Stephen Covey's *Seven Habits of Highly Effective People* (1989) very helpful. Three or four students have said this book changed their lives. Covey's book is also a perennial best-seller. The earlier you learn the skills and strategies for effective project work and teamwork, the more productive you will be and the easier life will be for you later. Start now.

Although the up-front goal of this book is to facilitate the development of teamwork and project management skills in engineering students, the deeper goals are to change the climate in engineering courses and programs from competitive or negative interdependence to cooperative or positive interdependence; from suspicion, mistrust, and minimal tolerance of others to acceptance, trust, and valuing others; from egocentric "What's in it for me?" to community "How are we doing?"; from a sense of individual isolation and alienation to a sense of belonging and acceptance. I recognize that these are lofty goals, but until we not only take responsibility for our own learning and development but also take more responsibility for the learning and development of others, we will not benefit from synergistic interaction.

If you find that teamwork and project management, and perhaps even leadership, are of great interest to you, then you may want to read some of the business magazines, such as *Business Week,* the *Harvard Business Review,* or my favorite, *Fast Company.* Check out your local bookseller or a bookstore on the Internet for some spare-time reading on these topics. You'll find an enormous literature available.

If you're more interested in teamwork and project management specifically within engineering and technology, then I suggest that you look into some of the books and video documentaries on projects, such as Karl Sabbagh's work—*Skyscraper* (1991) and *21st Century Jet* (1996).

Closing Reflection: On Reflection

Much of the material taught and concepts covered in college consist of *declarative knowledge,* which emphasizes knowing *that;* whereas the heart of project management consists of *procedural knowledge,* which emphasizes knowing *how.* This distinction between knowing that and knowing how was articulated by Ryle (1949). Furthermore, a lot of the essential procedural knowledge needed for success is implicit (or tacit, as Polanyi, 1958, 1966, described it) rather than explicit. This implicit, or "insider," knowledge is usually picked up on the job; however, I'm convinced that we can do a much better job of preparing people to "pick up" this knowledge. The opportunities for reflection I've encouraged you to engage in, as well as my own reflections, are designed to help you more quickly pick up the tacit dimension.

Best wishes to you in your project endeavors.

References

Covey, Stephen R. 1989. *The seven habits of highly effective people.* New York: Simon & Schuster.

Polanyi, Michael. 1958. *Personal knowledge: Towards a post-critical philosophy.* New York: Harper & Row.

———. 1966. *The tacit dimension.* New York: Doubleday.

Ryle, Gilbert. 1949. *The concept of mind.* London: Hutchinson.

Sabbagh, Karl. 1991. *Skyscraper: The making of a building.* New York: Penguin.

———. 1996. *Twenty-first century jet: The making and marketing of the Boeing 777.* New York: Scribner's.

Index